NATIONAL ACADEMIES *Sciences Engineering Medicine*

NATIONAL ACADEMIES PRESS
Washington, DC

A Vision for Continental-Scale Biology

Research Across Multiple Scales

Committee on Research at Multiple Scales:
A Vision for Continental Scale Biology

Board on Environmental Studies and Toxicology

Board on Life Sciences

Division on Earth and Life Studies

Consensus Study Report

NATIONAL ACADEMIES PRESS   500 Fifth Street, NW   Washington, DC 20001

This activity was supported by a contract between the National Academy of Sciences and the National Science Foundation (Contract No. AWD-000822). Any opinions, findings, conclusions, or recommendations expressed in this publication do not necessarily reflect the views of any organization or agency that provided support for the project.

International Standard Book Number-13: 978-0-309-71135-7
International Standard Book Number-10: 0-309-71135-5
Digital Object Identifier: https://doi.org/10.17226/27285
Library of Congress Control Number: 2025932609

This publication is available from the National Academies Press, 500 Fifth Street, NW, Keck 360, Washington, DC 20001; (800) 624-6242 or (202) 334-3313; http://www.nap.edu.

Copyright 2025 by the National Academy of Sciences. National Academies of Sciences, Engineering, and Medicine and National Academies Press and the graphical logos for each are all trademarks of the National Academy of Sciences. All rights reserved.

Printed in the United States of America.

Suggested citation: National Academies of Sciences, Engineering, and Medicine. 2025. *A Vision for Continental-Scale Biology: Research Across Multiple Scales*. Washington, DC: The National Academies Press. https://doi.org/10.17226/27285.

The **National Academy of Sciences** was established in 1863 by an Act of Congress, signed by President Lincoln, as a private, nongovernmental institution to advise the nation on issues related to science and technology. Members are elected by their peers for outstanding contributions to research. Dr. Marcia McNutt is president.

The **National Academy of Engineering** was established in 1964 under the charter of the National Academy of Sciences to bring the practices of engineering to advising the nation. Members are elected by their peers for extraordinary contributions to engineering. Dr. John L. Anderson is president.

The **National Academy of Medicine** (formerly the Institute of Medicine) was established in 1970 under the charter of the National Academy of Sciences to advise the nation on medical and health issues. Members are elected by their peers for distinguished contributions to medicine and health. Dr. Victor J. Dzau is president.

The three Academies work together as the **National Academies of Sciences, Engineering, and Medicine** to provide independent, objective analysis and advice to the nation and conduct other activities to solve complex problems and inform public policy decisions. The National Academies also encourage education and research, recognize outstanding contributions to knowledge, and increase public understanding in matters of science, engineering, and medicine.

Learn more about the National Academies of Sciences, Engineering, and Medicine at **www.nationalacademies.org**.

Consensus Study Reports published by the National Academies of Sciences, Engineering, and Medicine document the evidence-based consensus on the study's statement of task by an authoring committee of experts. Reports typically include findings, conclusions, and recommendations based on information gathered by the committee and the committee's deliberations. Each report has been subjected to a rigorous and independent peer-review process and it represents the position of the National Academies on the statement of task.

Proceedings published by the National Academies of Sciences, Engineering, and Medicine chronicle the presentations and discussions at a workshop, symposium, or other event convened by the National Academies. The statements and opinions contained in proceedings are those of the participants and are not endorsed by other participants, the planning committee, or the National Academies.

Rapid Expert Consultations published by the National Academies of Sciences, Engineering, and Medicine are authored by subject-matter experts on narrowly focused topics that can be supported by a body of evidence. The discussions contained in rapid expert consultations are considered those of the authors and do not contain policy recommendations. Rapid expert consultations are reviewed by the institution before release.

For information about other products and activities of the National Academies, please visit www.nationalacademies.org/about/whatwedo.

COMMITTEE ON RESEARCH AT MULTIPLE SCALES: A VISION FOR CONTINENTAL-SCALE BIOLOGY[1]

Members

JIANGUO LIU (*Chair*), Michigan State University
JEANNINE CAVENDER-BARES, University of Minnesota, Harvard University
BALA CHAUDHARY, Dartmouth College
BRIAN J. ENQUIST, University of Arizona; Santa Fe Institute
JACK A. GILBERT, University of California, San Diego
N. LOUISE GLASS, University of California, Berkeley
SCOTT GOETZ, Northern Arizona University
STEPHANIE E. HAMPTON, Carnegie Institution for Science
INÉS IBÁÑEZ, University of Michigan
CHELCY F. MINIAT, U.S. Department of Agriculture Forest Service
SHAHID NAEEM, Columbia University
PHOEBE L. ZARNETSKE, Michigan State University

Staff

CLIFFORD S. DUKE, Study Director
NATALIE ARMSTRONG, Program Officer
KAVITA BERGER, Director, Board on Life Sciences
LAYLA GARYK, Senior Program Assistant
DARLENE GROS, Senior Program Assistant
KATHERINE R. KANE, Senior Program Assistant
THOMASINA LYLES, Senior Program Assistant
TRISHA TUCHOLSKI, Program Officer
SABINA VADNAIS, Associate Program Officer
MAGGIE WALSER, Associate Executive Director

Sponsor

NATIONAL SCIENCE FOUNDATION

---

[1] All committee members serve as an individual rather than as a representative of a group or organization. The contributions of the committee members do not necessarily reflect the views of their employers or affiliated organizations.

BOARD ON ENVIRONMENTAL STUDIES AND TOXICOLOGY

Members

FRANK W. DAVIS (*Chair*), University of California, Santa Barbara
ANN M. BARTUSKA, U.S. Department of Agriculture (*until August 2024*)
DANA BOYD BARR, Emory University
WEIHSUEH A. CHIU, Texas A&M University
FRANCESCA DOMINICI, Harvard University
MAHMUD FAROOQUE, Arizona State University
MARIE C. FORTIN, Merck
MARIE L. MIRANDA, University of Illinois, Chicago
MELISSA J. PERRY, George Mason University
JOSHUA TEWKSBURY, Smithsonian Tropical Research Institute (*until August 2024*)
SACOBY M. WILSON, University of Maryland
TRACEY J. WOODRUFF, University of California, San Francisco

Staff

CLIFFORD S. DUKE, Board Director
NATALIE ARMSTRONG, Program Officer
LESLIE BEAUCHAMP, Senior Program Assistant
ANTHONY DEPINTO, Associate Program Officer
KATHRYN GUYTON, Senior Program Officer
KATHERINE R. KANE, Senior Program Assistant (*until March 2024*)
LAURA LLANOS, Finance Business Partner
THOMASINA LYLES, Senior Program Assistant
RAYMOND WASSEL, Scholar (*until June 2024*)

BOARD ON LIFE SCIENCES

Members

ANN M. ARVIN (*Chair*), Stanford University
DENISE N. BAKEN, Shield Analysis Technology, LLC
TANYA Y. BERGER-WOLF, Ohio State University
VALERIE H. BONHAM, Kennedy Krieger Institute
PATRICK M. BOYLE, Ginkgo Bioworks
DOMINIQUE BROSSARD, University of Wisconsin–Madison
MAURO COSTA-MATTIOLI, Altos Labs; Baylor College of Medicine
GERALD L. EPSTEIN, Johns Hopkins University Center for Health Security
INDIA G. HOOK-BARNARD, Engineering Biology Research Consortium
BERONDA MONTGOMERY, Grinnell College
LOUIS J. MUGLIA, Burroughs Wellcome Fund
ROBERT NEWMAN, AMP Health
LUCILA OHNO-MACHADO, Yale University
SUDIP S. PARIKH, American Association for the Advancement of Science
NATHAN D. PRICE, Thorne HealthTech
SUSAN R. SINGER, St. Olaf College
DAVID R. WALT, Harvard Medical School
PHYLLIS M. WISE, Colorado Longitudinal Study

Staff

KAVITA BERGER, Board Director
ANDREW BREMER, Program Officer
JESSICA DE MOUY, Research Associate
CYNTHIA GETNER, Finance Business Partner
LAYLA GARYK, Senior Program Assistant
NIA D. JOHNSON, Senior Program Officer
LYLY LUHACHACK, Program Officer
DASIA McKOY, Senior Program Assistant
CHRISTL SAUNDERS, Program Coordinator
AUDREY THEVENON, Senior Program Officer
TRISHA TUCHOLSKI, Program Officer
SABINA VADNAIS, Associate Program Officer
NAM VU, Senior Program Assistant

Reviewers

This Consensus Study Report was reviewed in draft form by individuals chosen for their diverse perspectives and technical expertise. The purpose of this independent review is to provide candid and critical comments that will assist the National Academies of Sciences, Engineering, and Medicine in making each published report as sound as possible and to ensure that it meets the institutional standards for quality, objectivity, evidence, and responsiveness to the study charge. The review comments and draft manuscript remain confidential to protect the integrity of the deliberative process.

We thank the following individuals for their review of this report:

ALAN HASTINGS, University of California, Davis
JAMES HEFFERNAN, Duke University
ALAN KNAPP, Colorado State University
DANIEL PARK, Purdue University
DAVID SCHIMEL, NASA Jet Propulsion Laboratory, Caltech
SUSAN RUNDELL SINGER, St. Olaf College
PATRICIA SORANNO, Michigan State University
MALAK TFAILY, University of Arizona

Although the reviewers listed above provided many constructive comments and suggestions, they were not asked to endorse the conclusions or recommendations of this report nor did they see the final draft before its release. The review of this report was overseen by **KATHY WEATHERS,** Cary Institute of Ecosystem Studies, and **MARY POWER (NAS),** University of California, Berkeley. They were responsible for making certain that an independent examination of this report was carried out in accordance with the standards of the National Academies and that all review comments were carefully considered. Responsibility for the final content rests entirely with the authoring committee and the National Academies.

Acknowledgments

Many people were essential in helping the committee accomplish its charge. The committee gratefully acknowledges the participants at its information-gathering sessions, who provided insights and viewpoints pertinent to the committee's task (see Appendix B). We thank Todd Anderson, Katharina Dittmar, Anika Dzierlenga, Thomas Elmqvist, Emiley Eloe-Fadrosh, Noah Fierer, Janet Franklin, Scott Hagerthey, Simon Levin, Sydne Record, David Schimel, Woody Turner, Michael Wilson, and Marten Winter for their contributions at the committee's first information-gathering session. We thank Jennifer Balch, Matthew Barnes, Tanya Berger-Wolf, Jessica Ernakovich, Alan Hastings, James Heffernan, Matthew Jones, Christopher Kempes, Elena Litchman, Paula Mabee, Osvaldo Sala, Lisette de Senerpont Domi, Margaret Torn, Susan Trumbore, and Yaxing Wei for their contributions at the committee's second information-gathering session. Presenters at the committee's third information-gathering session were John Bargar, Sara Bombaci, Gillian Bowser, Rachel Buxton, Stephanie Russo Carroll, Theresa Crimmins, Cristina Eisenberg, Andrew Farnsworth, Nico Franz, Sarah Huebner, Danielle Ignace, Christopher Lepczyk, John Matsui, Bonnie McGill, Patrick Meyfroidt, Milton Newberry III, Brook Nunn, Daniel Park, Jesús Pinto-Ledezma, Charuleka Varadharajan, Christine Wilkinson, and Elise Zipkin. We thank Sara Goeking for input on the USDA Forest Service Forest Inventory and Analysis network and Gavin Jones for input on acoustic recording units.

In addition, we are grateful to the National Science Foundation (NSF) for sponsoring the study, and to the NSF staff, Matt Kane and Erwin Gianchandani, for their presentations to the committee.

The committee is grateful to the staff of the National Academies of Sciences, Engineering, and Medicine who contributed to producing this report, especially the outstanding and tireless study staff: Natalie Armstrong, Kavita Berger, Clifford Duke, Layla Garyk, Darlene Gros, Katherine Kane, Thomasina Lyles, Trisha Tucholski, and Sabina Vadnais. Thanks also go to the staff of the Division on Earth and Life Studies who provided additional support, including Elizabeth Eide, Lauren Everett, Nancy Huddleston, Radiah Rose, and Maggie Walser. This project also received important assistance from Laura Llanos (Office of Financial Administration). We also appreciate the editorial and graphics support of Solmaz Spence and Stacy Jannis, respectively.

Contents

ACRONYMS AND ABBREVIATIONS   xvii

SUMMARY   1

1 INTRODUCTION   15
Paving the Way for a Continental-Scale Biology, 16
Committee's Approach to the Statement of Task, 19
References, 22

2 THEMES FOR A CONTINENTAL-SCALE BIOLOGY   24
Biodiversity and Ecosystem Function, 25
Resilience and Vulnerability, 27
Connectivity, 29
Sustainability of Ecosystem Services, 33
Examples of Interworking of the Four Themes, 36
References, 43

3 THEORETICAL UNDERPINNINGS FOR A CONTINENTAL-SCALE BIOLOGY   49
Overview and Problem Statement, 49
Challenges in Developing CSB Theories, 53
Requirements of CSB Theories, 55
Challenges in Connecting Research Across Scales, 64
Conclusions on Developing Theory to Connect Research Across Scales, 67
References, 70

4 RESEARCH INFRASTRUCTURE THAT ENABLES CONTINENTAL-SCALE BIOLOGY   78
Introduction, 78
Tools, 78
Networks, 94
Centers of Synthesis, 106
Attributes of Successful Tools and Networks and Challenges, 107
Recommendation and Conclusion, 110
References, 112

5 TRAINING AND CAPACITY BUILDING TO ENABLE CONTINENTAL-SCALE BIOLOGY   122
Data Literacy Training and Capacity Building, 124
Interdisciplinary Team Science Training, 129
Challenges That Limit Effective Research Across Scales and Training Approaches to Overcome Them, 133
Evidence-Based Methods to Promote Diversity, Equity, Inclusion, and Accessibility in Continental-Scale Biology, 134
A Path Forward for Training and Capacity Building in Continental Biology, 139
References, 140

6 OVERARCHING RECOMMENDATIONS AND VISION   144
Overarching Recommendations, 144
Vision, 146
References, 147

APPENDIXES

A COMMITTEE MEMBER BIOGRAPHICAL SKETCHES   149

B PUBLIC MEETING AGENDAS   154

Boxes, Figures, and Tables

BOXES

1-1 Related NSF Initiatives, 17
1-2 Characteristics That Define Continental-Scale Biology, 20

2-1 Cataloging Earth's Microbial Biodiversity Across Multiple Habitats, 26
2-2 Biodiversity in Coral's Slimy Biofilms as an Indicator of Reef Resilience, 30
2-3 Effects of Local, Landscape, and Regional Drivers of Vulnerability to Plant Invasions, 32
2-4 A Metacoupling Framework to Help Optimize Salmonid Research, Management, and Policymaking, 34
2-5 Bridging Spatial Scales: How the Ecosystem Service of Slope Stability Changes with Rainfall Amounts and Forest Species Composition, 37
2-6 Effect of Tree Fecundity on the Long-Term Maintenance of Ecosystem Services Provided by Forested Areas, 38

3-1 Role of Theory in the Core Themes of CSB, 54
3-2 Developing, Testing, and Advancing Theory, 58

4-1 Connecting the Tools and Networks that Enable CSB to its Core Themes, 81
4-2 Application of Omics Technologies That Can Be Applied to CSB, 84
4-3 Application of Remote Sensing to Collect Vegetation Spectral and Structural Data Relevant to CSB, 89
4-4 The Nutrient Network, 104

5-1   Connecting Core Themes to Training and Capacity-Building Efforts Used in Research Across Scales, 125
5-2   Examples of Training and Capacity-Building Efforts Used to Successfully Connect Research Across Scales, 126

FIGURES

S-1   CSB provides a framework for addressing environmental and societal challenges that cut across biological organizational scales from molecule to cells and tissues, organisms to populations and communities, ecosystems to the biome to the biosphere, 3
S-2   Relationships among the four themes, 5

1-1   Biological organizational scales of continental-scale biology, 21

2-1   Relationships among the four themes, 25
2-2-1 Diverse microbial communities make up coral reef biofilms, 31
2-4-1 A metacoupling framework provides a tool for studying the interactions between humans and nature, 35
2-2   Elements of the Comprehensive Everglades Restoration Plan, 39
2-3   Research on pandas, people, and policies across scales from DNA to planet Earth, 41

3-1-1 Near-term ecological forecasting cycle, 59
3-2   Hypothetical biodiversity curves illustrating different scenarios and impacts on biodiversity targets, 69

4-1a Relationships among tools, networks, and synthesis centers to biological processes in the context of biological knowledge, 79
4-1b Temporal and spatial scales of biological and physical processes and patterns in the context of multiscale biology, 80
4-2   National Ecological Observatory Network, 82
4-3-1 Image of airborne platform using full-range spectroscopy (400–2,500 nm) and remotely sensed lidar, 89
4-4-1 Nutrient Network site locations, 105

5-1   Numerous current and former training programs exist to support a scientific workforce, 123

TABLES

4-1   Scientific Observational and Experimental Networks, 95

Acronyms and Abbreviations

ABoVE	Arctic-Boreal Vulnerability Experiment
AI	artificial intelligence
ARU	acoustic recording unit
BIEN	Botanical Information and Ecology Network
BOREAS	Boreal Ecosystem-Atmosphere Study
BRC-BIO	Building Research Capacity of New Faculty in Biology
CERP	Comprehensive Everglades Restoration Plan
CHANS	coupled human and natural systems
CIMER	Center for the Improvement of Mentored Experiences in Research
CSB	continental-scale biology
DAAC	Distributed Active Archive Center
DEI	diversity, equity, and inclusion
DroughtNet	Drought Network
eDNA	environmental deoxyribonucleic acid
EMBeRS	Employing Model-Based Reasoning in Socio-Environmental Synthesis
eRNA	environmental ribonucleic acid
ESIIL	Environmental Data Science Innovation & Inclusion Lab
ESM	Earth system model
EWS	early warning signals

FAIR	findable, accessible, interoperable, and reusable
FIA	Forest Inventory and Analysis
GBIF	Global Biodiversity Information Facility
GEDI	Global Ecosystem Dynamics Investigation
GLEON	Global Lake Ecological Observatory Network
HHMI	Howard Hughes Medical Institute
ICARUS	International Cooperation for Animal Research Using Space
ISLSCP	International Satellite Land Surface Climatology Project
ITEK	Indigenous and Traditional Ecological Knowledge
LBA	Large-scale Biosphere-Atmosphere Experiment
lidar	light detection and ranging
LTER	Long-Term Ecological Research Network
ML	machine learning
MS	mass spectrometry
MST	metabolic scaling theory
NADP	National Atmospheric Deposition Program
NASEM	National Academies of Sciences, Engineering, and Medicine
NCEAS	National Center for Ecological Analysis and Synthesis
NEON	National Ecological Observatory Network
NGO	nongovernmental organization
NPN	National Phenology Network
NSF	National Science Foundation
NTB	neutral theory of biodiversity
NutNet	Nutrient Network
OSTP	Office of Science and Technology Policy
PI	principal investigator
RCN	Research Coordination Network
RFP	request for proposals
rRNA	ribosomal ribonucleic acid
SBG	Surface Biology and Geology
SESYNC	Socio-Environmental Synthesis Center
STEMM	science, technology, engineering, mathematics, and medicine
TDT	trait driver theory
TRY	Plant Trait Database
USACE	U.S. Army Corps of Engineers

Summary[1]

INTRODUCTION

Our planet is facing many complex environmental challenges, including biodiversity loss and rapidly changing climate conditions, driven by intensifying human–nature interactions worldwide. Human actions cause global and regional changes, having profound impacts at local scales. Conversely, local-scale environmental changes can contribute to regional and global impacts. For example, the spread of invasive species is a global phenomenon driven primarily by trade and other human activities and can involve organisms traveling many thousands of miles to a new location. However, local-scale data on ecosystem dynamics and on how new organisms may adapt can help to guide the most appropriate responses to limit their spread. Similarly, at global and regional scales, human-caused climate change can lead to environmental shifts such as more frequent droughts that may impact the health of forests; conversely, forests may be able to mitigate the effects of climate change via carbon sequestration. Further, local changes in vegetation may drive local to regional changes in atmospheric circulation and create "ecoclimate teleconnections" over even larger scales.

Increasingly, scientists are recognizing that research across multiple scales, from the molecular to regional to global, can provide new insights into the interacting factors that are contributing to these challenges. Dramatic advances in the biological sciences in recent years mean that researchers now have some of the tools needed to study life at many scales, from identifying mutations in a single gene to monitoring changes in plants, animals, and microbes over an entire continent. Available tools include networked observatories of standardized biological sampling across ecoclimatic gradients; experiments that manipulate variables and are replicated across space and/or time;

---

[1] This summary does not include references. Citations for the information presented herein are provided in the main text.

observational studies using remote sensing or sensor technology to capture population-level or community dynamics; genetic and microbial sampling; biodiversity collections; and modeling approaches enabling inferences from observations or predicting outcomes over large spatial extents and through time. These tools have the potential to usher in a new era of continental-scale biology (CSB) in which researchers use new multiscaled, multidisciplinary theory, advances in research infrastructure, and the development of a skilled workforce to address challenges that cross multiple scales from *molecules to organisms, and from ecosystems to biomes to the biosphere (Figure S-1)*. CSB offers the wide lens needed to address the urgent challenges of declining biodiversity, climate change, emerging infectious diseases, the spread of invasive species, food security, and environmental justice. CSB is an interdisciplinary frontier that will require theoretical, empirical, and cultural integration across allied disciplines as varied as hydrology, engineering, and social sciences.

STATEMENT OF TASK

As the impacts of climate change, biodiversity loss, and other stressors accelerate, there is an urgent need to gain knowledge of these critical factors, how they interact, and how they should inform decision making. Several recent National Science Foundation (NSF) initiatives—for example, Reintegrating Biology, Understanding the Rules of Life, the Biological Integration Institutes, and Macrosystems Biology—have sought to enhance our understanding of biological systems by integrating methods and knowledge from the many subdisciplines of biology and other scientific disciplines, and at many different scales. This report, prepared at the request of NSF, complements these initiatives by identifying productive routes for the development of continental, multiscale biology and strategies to facilitate the concomitant reunification of biology across organizational, spatial, and temporal scales.

The statement of task for the report was as follows:

> An ad hoc committee of the National Academies of Sciences, Engineering, and Medicine will conduct a consensus study to identify how biological research at multiple scales can inform the development of a continental scale biology. The committee will convene a series of virtual community workshops to inform its deliberations.
>
> Specifically, the committee will identify and discuss:
>
> - Practices that have been used successfully to translate knowledge, approaches, and tools from small-scale biological research to regional- and continental-scale, and vice versa. For example, how might understanding fine-scale biological discoveries alter specific measurements or experiments that researchers undertake at larger scales and vice versa?
> - Challenges that prevent uptake of these practices.
> - Specific research questions that could serve as pilots for implementing research projects that integrate one or more successful practices.
>
> Finally, the committee will review and refine the practices and questions into a set of recommendations for the research community, funders, and decision makers.

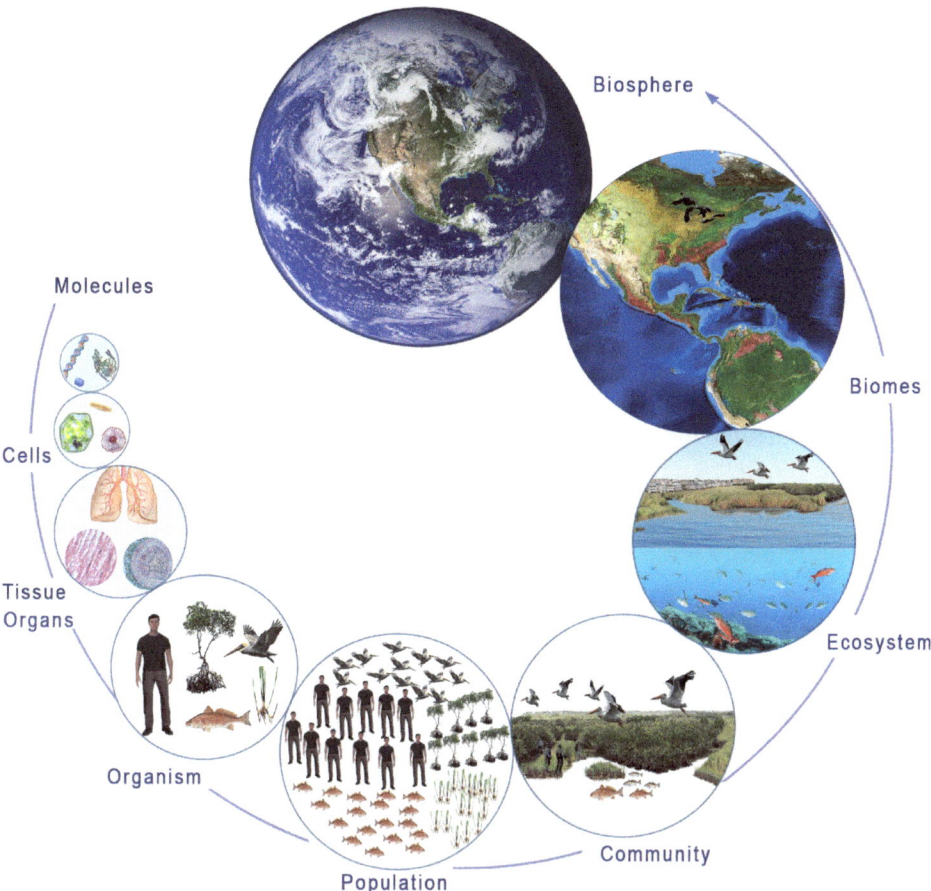

FIGURE S-1 CSB provides a framework for addressing environmental and societal challenges that cut across biological organizational scales from molecule to cells and tissues, organisms to populations and communities, ecosystems to the biome to the biosphere.
SOURCE: Stacy Jannis.

COMMITTEE'S APPROACH TO THE STATEMENT OF TASK

The committee approached the statement of task by first identifying key characteristics that define CSB. The report envisions a CSB that addresses biological processes and patterns that emerge at broad *organizational*, *spatial*, and/or *temporal scales* that cannot be answered by observations and experiments conducted at either fine or large scales alone. CSB inherently incorporates *multiple scales*, from the subcellular to the global biosphere, from local to global spatial extents, from less than a second to millennia in time. Specific CSB research may operate across one, two, or all three kinds of scales:

organizational, spatial, or temporal. Further, CSB takes a systems approach, treating biological systems as part of coupled human and natural systems, given widespread human impacts and intensifying human–nature interactions worldwide. Understanding scalability and how biological properties (e.g., patterns and processes) vary or remain the same across scales is an important area of inquiry for CSB. CSB is enabled by emerging theory; recent developments of experimental and observational networks, tools, and analytical techniques; and changes in the culture of biological science that facilitate collaboration among multidisciplinary teams with members from around the globe.

The committee identified four integrated major research themes that CSB is particularly well positioned to support and for each theme identified potential questions that could serve as pilots for implementing research projects. These themes and questions are neither mutually exclusive nor comprehensive, but rather are intended to stimulate discussion about directions for the development of the field. Following the themes, the committee developed conclusions and recommendations on theory, research infrastructure, and training and capacity-building efforts to develop and maintain CSB; overarching recommendations; and a vision for CSB.

THEMES AND POTENTIAL RESEARCH QUESTIONS

The committee identified four major themes in the types of research that CSB is particularly well positioned to support and, in each theme, identified a series of research questions that could serve as pilots for implementing research projects. The four major research themes are interrelated, as shown in Figure S-2. For instance, the fourth theme, "sustainability of ecosystem services" is related to the first three because biodiversity and ecosystem functions are essential to produce ecosystem services, resilience and vulnerability are key in maintaining ecosystem services, and connectivity is central to shaping distribution and use of ecosystem services across space and over time.

Biodiversity and Ecosystem Function

A pivotal theme of biological research over the past half century has been the relationship between biological variation, or biodiversity (in terms of taxonomy, function, phenotype, genotype, or phylogenetic placement) and ecosystem functions, for example, the cycling of carbon, nitrogen, phosphorus, sulfur, oxygen, water, and trophic interactions. Much of the experimental work has been conducted at local scales, but generalities have emerged to indicate that biodiversity is important to ecosystem function at multiple scales. As yet, there is scant evidence available to define these relationships and to decipher how they may change across scales—information that will be central to determining how biodiversity across spatial and temporal scales drives local-to-Earth system function. A CSB approach offers the opportunity to use emerging tools, including satellite remote sensing, genetic sequencing, multi-omics, artificial intelligence, and automated monitoring systems, together with developing theory, to examine the

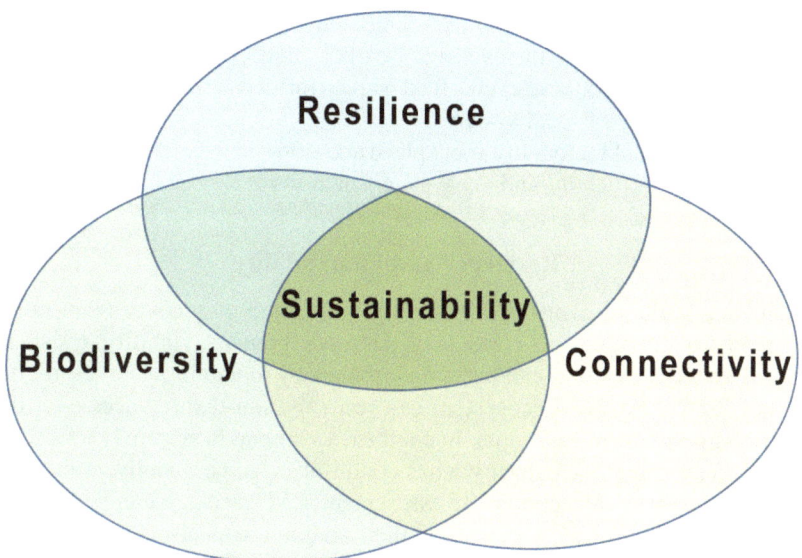

FIGURE S-2 Relationships among the four themes. Sustainability lies at the center, incorporating elements of biodiversity, resilience, and connectivity. The latter three themes overlap with each other while also having their own domains of inquiry.
SOURCE: Stacy Jannis.

linkages between biodiversity and ecosystem functions across scales, and to learn how these relationships may shift in the face of global change. Potential research questions related to this theme include:

- How does complexity of a biological system at one scale influence emergent properties of the system at the next scale, and how do those properties feed back to influence complexity and variation at all scales?
- How do ecosystem functions emerge from biological complexity, what mechanisms are involved, and how do these vary across temporal and spatial scales? How do major global changes—including climate change, human-induced land- and water-use changes—and the spread of pests and pathogens, impact how the diversity of life influences and interacts with functions and processes of emergent systems?
- How do human-induced changes in land use and water use, habitat fragmentation, and, conversely, habitat connectivity at local and regional scales affect continental-scale biodiversity?
- How does theory support our understanding of the causes and consequences of continental-scale biodiversity in the context of major global changes and human activities?

- How does the distribution of composition, diversity, and complexity of ecosystems influence the cycling of carbon, nutrients, and water at continental scales, and how do these cycles feed back to influence the composition, diversity, and complexity of local systems?
- How do internal factors in a focal place and flows between focal and other places influence biodiversity and ecosystem functions across scales?

Resilience and Vulnerability

Resilience is the capacity of a system to withstand or recover from human and environmental disturbances. It consists of three components: stability, resistance, and recovery. A stable ecosystem can resist disturbance by maintaining structure, function, and composition. A resilient ecosystem can recover from disturbances over time and essentially return to its original state. In contrast, a vulnerable system is sensitive to disturbance and lacks adaptive capacity when conditions change. Ideally, multiple metrics should be used to provide a composite assessment. CSB research can be used to assess the sensitivity of biological systems and their adaptive capacity to resist, recover, or change in biological composition and function in response to disturbances. Potential research questions related to this theme include:

- How do biodiversity and human activities influence the resilience of biological systems at multiple spatial and temporal scales? How does this relationship shift across scales of biological organization, from the population (diversity within a species) to the biome?
- Which ecosystems and functions at what scales are most exposed and vulnerable to natural and human disturbances?
- What are the tipping points and thresholds of resilience and vulnerability across scales? What continental-scale drivers affect the spatial and temporal patterns of ecosystem vulnerability at local, regional, and national scales, and how do local to regional drivers affect vulnerability and resilience at the continental scale?
- How does the resilience or vulnerability of one system, for example, to land use or climate change, affect resilience or vulnerability of other systems near and far?

Connectivity

Relationships between habitats and living things in one place can have profound effects on ecosystems in other places, both near and far. However, more systematic ways are needed to effectively integrate the influences of biota, habitats, ecosystem functions and materials, abiotic components (e.g., air, climate, geophysical conditions, soil/land, water), and humans across time and space. CSB can offer tools that can combine data and insights from each of these realms across time and space, helping to build a stronger understanding of the system as a whole. For example, interdisciplinary frameworks provide a systematic way to combine data and insights from different ecosystems across time and space. Potential research questions related to this theme include:

- How do human–nature interactions vary across organizational, spatial, and temporal scales, and what are the impacts of economic globalization and environmental changes on these dynamics?
- How do human activities and environmental changes (including climate change) in one system generate cascading effects across adjacent and distant systems?
- How do interactions within a focal system, between adjacent systems, and between distant systems create feedback, synergistic, or trade-off effects on resilience and sustainability of biodiversity and ecosystems across scales?
- How do risks associated with invasive species, pathogen emergence, and other shocks such as natural disasters spread across adjacent and distant systems at multiple scales?

Sustainability of Ecosystem Services

Ecosystem services are the benefits people obtain from ecosystems, including the provisioning of food and water; the regulation of floods, drought, land degradation, and disease; nutrient cycling and soil formation; and nonmaterial benefits such as recreational, spiritual, and religious benefits of nature. Sustainability relates to an ecosystem's ability to continue to provide services, without any change in the level of service being provided, in the face of environmental disturbances or state changes. CSB research brings knowledge, via systems' resilience and vulnerability to stressors, on how sustainability of ecosystem services may be affected across temporal and spatial scales. Potential research questions related to this theme include:

- Which ecosystem services are most sustainable at different scales in the face of human activities and environmental change?
- How do socioeconomic, urbanization, and land-use changes affect ecosystem services across local to continental scales?
- What factors across scales affect ecosystem services at the scale of interest, and how does the sustainability of an ecosystem service shift across spatial scales?
- How does sustainability of ecosystem services at one scale affect sustainability at other scales?

CONCLUSIONS AND RECOMMENDATIONS ON THEORY, RESEARCH INFRASTRUCTURE, AND TRAINING AND CAPACITY BUILDING TO DEVELOP CSB

Theory

Theory is essential to CSB because it provides a structured framework that allows us to understand and explain biological phenomena. Theory helps bridge gaps in our current understanding of both small- and large-scale biological and ecological processes and patterns. Requirements of CSB theories are several: (1) they need to be applicable at multiple spatial and temporal scales, encompassing attributes of individuals (traits, genes), populations, and species assemblages on landscapes up to the entire biosphere; (2) they need to mesh with diverse datasets, technologies, and monitoring programs, both guiding their development and being flexible enough to change in response to new knowledge; (3) they must forge connections across biological processes from the molecular and cellular levels to populations, entire ecosystems, and the biosphere; and (4) they should be comprehensive, unifying disparate domains from microbial processes to ecological and physiological processes and material flux through the biosphere. The committee concluded that theory is especially needed in three areas.

Conclusion 3-1: Theory is needed that links research at multiple organizational, spatial, and temporal scales, from micro to meter to landscapes up to the biosphere. The multidimensional and hierarchical multiscale nature of biodiversity requires solutions that can address cross-scale questions and identify cross-scale phenomena. For this approach, theory is needed that meshes with our current technologies and informatics that collectively monitor biosphere processes. Theory also needs to be based on conceptual frameworks that integrate multiscale data. This includes molecular, microbial processes, genomes, environmental DNA, metagenomics, metatranscriptomics, stable isotope labeling, and metabolomics. These data sources are crucial for linking local ecological and physiological processes of organisms to broader patterns and data collection efforts such as the distribution of species, movement of individuals and species, the functioning of ecosystems, and the flux of material and matter through the biosphere at multiple scales. This integration will enable the refinement and development of CSB theory, enhancing our ability to model and manage environmental changes effectively. By incorporating larger-scale data from remote sensing, tower-based systems, global animal tracking, and sensor networks, we can enrich this framework, providing a more comprehensive understanding necessary for predictive modeling and sustainable ecosystem management.

Conclusion 3-2: Theory is needed to improve climate and global change models by including biological feedbacks. Biological processes that result in feedbacks to ecosystems and climate are a challenge to incorporate into climate and global change theories, presenting considerable uncertainty. The inclusion of biological feedback in theories of continental-scale models of global change will enhance our ability to predict future trends and identify cross-scale solutions and will be a key component of clarifying and improving climate and global change models. Refinement of biological feedback theories into continental-scale models and extension to climate and global change theories will improve our ability to predict future trends as well as identify solutions that cut across scales.

Conclusion 3-3: Theory is needed that incorporates the effects of human-induced environmental changes (including climate change) to predict changes within an ecosystem and to assess metacoupled cascading effects across adjacent and distant systems. Theory is needed to predict interactions among system components across all scales that impact adjacent and distant environments. The inclusion of theory that incorporates human activity will enable the prediction of synergistic, cascading, or trade-off effects on resilience and sustainability of ecosystems and the biosphere across time and space.

Research Infrastructure

Fully responding to the challenges of developing CSB requires both the enhancement of existing infrastructure and the development of new infrastructure. The committee offers the following recommendation and conclusion to NSF and other agencies.

Recommendation 4-1: To provide infrastructure specifically aimed at supporting continental-scale biology (CSB), the National Science Foundation (NSF) should consider the following options, as available resources permit:

- Explore the development of artificial intelligence (AI) and informatics tools, and open-access databases explicitly focused on CSB, synthesizing knowledge across scales, that would synergize with the synthesis work currently conducted at the Environmental Data Science Innovation & Inclusion Lab. A request for proposals (RFP) to support virtual infrastructure and computational science innovations would be of great value to realize the potential of CSB data. For example, linking remotely sensed spectral and structural measurements to physical and biological measurements on the ground could be advanced by developing new algorithmic modalities.
- Build new sensor modalities to improve data collection. This could be achieved by developing interdisciplinary funding opportunities that unite ecology, engineering, atmospheric science, remote sensing, hydrology, and other disciplines such as social sciences.
- Allocate resources for next-generation digitization of biodiversity collections to enhance their utility as reference standards for CSB and to enable the development of digital ecosystem twins. This will require new bioinformatics tools to enable access to and management of preserved and living collections to facilitate their utility for interpretation of in situ and remotely sensed data.
- Develop communities that can leverage interdisciplinary data from NSF platforms and various networks, akin to the use of National Ecological Observatory Network (NEON) data by researchers funded by the previous Macrosystem Biology program. An example would be incorporating macrosystems/synthesis research to create living data products (those that are continually updated) that inform biological processes at continental scale. If done, data from one platform could serve as calibration/validation for other data products and layers from other platforms to facilitate interpolation, extrapolation, and/or imputation. Major government assessments, such as the National Nature Assessment, could also leverage data provided by integration of these platforms.
- Support efforts to understand how to sample for continental-scale biological questions. Few spatially distributed networks with standardized sampling exist, and those that do exist require great resources. Investment is needed in research to understand sampling theory (time and space) for capturing continental scales and cross-boundary interactions.

- Explore joint support of integrative science via interagency (e.g., NASA, U.S. Department of Energy) RFPs, for example, multiscale coordinated interdisciplinary field campaigns (e.g., NASA Arctic-Boreal Vulnerability Experiment, the Biodiversity Survey of the Cape, and the Large-scale Biosphere-Atmosphere Experiment in Amazonia).
- Explore development of interagency incentives and mechanisms for public–private partnerships that can facilitate targeted private investment in data development and integration across scales and types, for example, engaging in AI-driven data analysis and data product development.

Conclusion 4-1: Development of research infrastructure for CSB would also benefit from actions by other agencies. Examples include the following:

- The Small Business Administration could develop agency-specific innovation research (Small Business Innovation Research) and technology transfer research (Small Business Technology Transfer Research) RFPs that focus on AI, machine learning, and sensor development for the biological and environmental sciences.
- NASA's continued support for the Surface Biology and Geology (SBG) mission, the only satellite instrument dedicated to biological processes that will specifically enable CSB, is an important contribution. SBG will provide continuous data across continents and the globe to fill in the gaps from NEON and enable baseline information to track changes in biological processes through time. Other NASA Explorer and Incubator missions recommended by the Decadal Survey for Earth Science and Applications from Space include lidar and radar measurements of ecosystem three-dimensional structure.
- Agencies engaged in these efforts could continue to support scientific assessments and action-oriented efforts that inform policies guided by or consistent with the UN Convention on Biological Diversity. These include the Global Biodiversity Information Facility, the Group on Earth Observations Biodiversity Observation Network, and the Intergovernmental Science-Policy Platform on Biodiversity and Ecosystem Services.
- Continue support of related domestic efforts that contribute to CSB, for example, the ongoing National Nature Assessment, U.S. Geological Survey national Biodiversity and Climate Change Assessment, and the U.S. contribution to the 30×30 Conservation Initiative.

Training and Capacity Building

Recommendation 5-1: The three key areas of training that funders, researchers, and educators should prioritize for developing a scientific workforce with the knowledge and skills necessary to address future challenges in CSB are data literacy, interdisciplinary team science, and promoting diversity, equity, inclusion, and accessibility.

These are not unique to CSB, but effective development of this field is particularly dependent upon them. Large-scale spatial and temporal data and data across scales are essential in CSB research; thus, data literacy—from basic to high-level expertise—will be necessary across team members to ensure an efficient workflow. Teamwork involving several disciplinary expertise and skill sets is also inherent in CSB research, such that effective communication and productive interactions across team participants with different backgrounds will be critical to ensure successful project outcomes. Additionally, addressing systemic issues such as career tracks and incentives is crucial for retaining skilled individuals in academia, which in turn maximizes creativity and productivity. To maximize creativity and productivity, team science for CSB also requires inclusivity and diverse perspectives.

OVERARCHING RECOMMENDATIONS

Encompassing all four themes discussed in this report is the fact that CSB is built on connections: from ecoclimatic teleconnections (causal links between phenomena in geographically distant regions), to feedbacks between ecosystems and ecosystem components, to cross-scale interactions that occur when phenomena at one organizational, temporal, or spatial scale influence another. In addition, virtually every natural system on Earth influences and is influenced by human activities, even over long distances. Bringing together these factors is a central challenge of CSB research. In response, the committee makes two overarching recommendations that will help to meet this challenge and support the development of the emerging field of CSB.

Overarching Recommendation 1: The National Science Foundation should establish a core program on CSB.

The committee recommends that NSF establish a new core program on CSB. As described previously, several recent NSF initiatives have sought to enhance understanding of biological systems by integrating the methods and knowledge from the many scientific subdisciplines and at many different scales. However, the scope of CSB is broader than that of any existing NSF program. For example, the core programs in the Division of Environmental Biology support "research and training on evolutionary and ecological processes acting at the level of populations, species, communities, and ecosystems," but CSB also addresses processes below population levels (e.g., subcellular, cellular) and above ecosystem levels (e.g., regional, continental).

Therefore, CSB would strongly benefit from the establishment of an NSF core program that provides a stable and dedicated funding source to support research addressing the interplay of multiple organizational, spatial, and temporal scales and is based on integrated yet flexible frameworks. This could be a joint effort among the relevant NSF divisions and directorates to help facilitate collaborations, both between the Division of Environmental Biology and other divisions within the Directorate for Biological Sciences (e.g., the Division of Integrative Organismal Systems and the Division of Molecular and Cellular Biosciences), and with other directorates, such as Mathematical and Physical Sciences; Computer and Information Science and Engineering, Office of Advanced Cyberinfrastructure; Engineering; and Social, Behavioral, and Economic Sciences. For example, collaboration with the Directorate for Technology, Innovation, and Partnerships could help with the development of new technologies that would advance CSB; work with the Directorate for Social, Behavioral, and Economic Sciences would help provide additional insight on the increasing influence of human activities on biological systems, and, conversely, the effects of biological systems on human well-being; and collaboration with the Directorate for Geosciences would support work on the linkages between geophysical and atmospheric processes and CSB.

Overarching Recommendation 2: Researchers and funders should develop CSB under integrated yet flexible frameworks.

As discussed in the connectivity theme, CSB addresses questions about biological processes and patterns that emerge at broad organizational, spatial, and/or temporal scales and treats biological systems as part of coupled human and natural systems, given widespread human impacts and intensifying human–nature interactions worldwide. Integrated yet flexible frameworks for CSB would enable researchers to better understand and contextualize the connections among the biological, abiotic, and socioeconomic realms, and the interactions within, between, and among adjacent and distant locations. Such frameworks would help researchers gain a holistic view of local- and regional-scale ecosystems and continental-scale environmental shifts—insights that will allow the development of more effective and sustainable solutions to the environmental and ecological challenges facing our planet. An example is metacoupling framework, which has been applied to analyses of many topics, including ecosystem services, resilience, vulnerability, biodiversity conservation, biogeochemical flows, climate change, freshwater use, land use, pollution, impacts of food imports on food security, and effects of international trade on deforestation. These and other potential frameworks to be developed would provide a sound foundation for future CSB research.

VISION FOR CONTINENTAL-SCALE BIOLOGY

Bold initiatives are needed to create a truly continental-scale biology that addresses questions across multiple organizational, spatial, and temporal scales. The vision for a new era of CSB involves integrating biological subdisciplines as well as other disciplines and scales of research to leverage the biological data revolution, addressing questions unanswerable by fine- or large-scale observations alone. This multidisciplinary systems approach can provide a comprehensive understanding of complex biological systems and their interactions with human and environmental factors, fostering innovative solutions to global challenges.

Although much progress has been made, there are many major gaps in knowledge, theory, data, networks, tools, and training and capacity building needed to support the vision for CSB. Filling these gaps will require the development of new theories and technologies that encompass not just biology, but atmospheric sciences, mathematics, engineering, physics, geosciences, environmental chemistry, and social sciences. Such effort is crucial to enhance the fundamental understanding of ongoing changes in biodiversity, ecosystem services, climate, disease spread, species invasion, gene flows, and biotic interactions. It is also needed to build workforce capacity by mentoring a new generation of innovative scholars and engaging leaders for global sustainability. By addressing these challenges with coordinated and innovative efforts, we can pave the way for a sustainable and resilient future, ensuring the well-being of our planet and its ecosystems.

1

Introduction

Our planet is facing many complex environmental challenges, from the loss of biodiversity to the effects of rapidly changing climate conditions—many of them driven by intensifying human–nature interactions worldwide. Human actions are causing global and regional changes that are having profound impacts at local scales, and conversely, environmental changes at local scales can contribute to regional and global impacts. For example, the spread of invasive species is a global phenomenon driven primarily by trade and other human activities and can involve organisms traveling many thousands of miles to a new location. However, local-scale data on ecosystem dynamics and on how the new organisms may adapt can help to guide the most appropriate responses to limit their spread (LaRue et al. 2021). Similarly, at global and regional scales, human-caused climate change can lead to environmental shifts such as more frequent droughts that may impact the health of forests; conversely, forests may be able to mitigate the effects of climate change via carbon sequestration (Bonan 2008). Further, local changes in vegetation may drive local to regional changes in atmospheric circulation and create "ecoclimate teleconnections" over even larger scales (Stark et al. 2016).

Increasingly, scientists are recognizing that research across multiple scales, from the molecular to regional to global, can provide new insights into the interacting factors that are contributing to these challenges (Heffernan et al. 2014). Dramatic advances in the biological sciences in recent years mean that researchers now have some of the tools needed to study life at many scales, from identifying mutations in a single gene to monitoring changes in plants, animals, and microbes over an entire continent. Available tools include networked observatories of standardized biological sampling across ecoclimatic gradients, experiments that manipulate variables and are replicated across space and/or time; observational studies that use remote sensing or sensor technology to capture population-level or community dynamics, genetic and microbial sampling, and biodiversity collections; and modeling approaches that connect disparate information to enable inferences from observations or predict outcomes over large spatial extents and through time.

A central challenge is to bring large volumes of new and existing data and information together in a way that improves the ability to classify, interpret, and predict biological and physical processes. The federal government has developed a number of networks that collect data and information to address a basic question or to respond to a societal concern. For example, the question of how and where acid deposition is affecting air and water quality is being investigated through the National Atmospheric Deposition Program and the Clean Air Status and Trends Network. Investigation of the biospheric fluxes of energy, carbon, and water across the U.S.' terrestrial ecosystems' boundary layer is being conducted by the U.S. Department of Energy's Ameriflux network. Some networks are established not to answer a particular question but rather function more as data collection or observational networks. For example, the National Ecological Observatory Network (NEON) collects long-term open-access ecological data in 81 field sites across the United States (see more on NEON in Chapter 4).

In addition to such networks, centers that facilitate synthesis by groups of scientists with diverse, but complementary expertise play useful roles in addressing questions of critical importance. Synthesis centers have been extraordinarily successful in bringing together scientists from different perspectives and skill sets to address specific challenges. NSF-funded synthesis centers in the United States have included the National Center for Ecological Analysis and Synthesis, the Socio-Environmental Synthesis Center, the National Institute for Mathematical and Biological Synthesis, the National Evolutionary Synthesis Center, and now the Environmental Data Science Innovation & Inclusion Lab.

PAVING THE WAY FOR A CONTINENTAL-SCALE BIOLOGY

As the impacts of climate change, biodiversity loss, and other stressors accelerate, there is an urgent need to gain knowledge of these critical factors, how they interact, and how they should inform decision making. Conducted at the request of NSF, this report identifies productive routes for the development of continental, multiscale biology and strategies to facilitate the concomitant reunification of biology across organizational, spatial, and temporal scales. The report complements recent NSF initiatives, for example, Reintegrating Biology, Understanding the Rules of Life, the Biological Integration Institutes, and Macrosystems Biology (Box 1-1), which have sought to enhance our understanding of biological systems by integrating methods and knowledge from the many subdisciplines of biology and other scientific disciplines, and at many different scales.

By integrating biological research across subdisciplines and across organizational, spatial, and temporal scales, continental-scale biology (CSB) promises to reveal new insights on complex biological phenomena and help inform responses to the planet's most pressing environmental crises. CSB seeks to enhance our understanding of cross-scale interactions across biological organizational, spatial, and temporal scales, recognizing that there are biological phenomena that may only be revealed across scales and

> **BOX 1-1**
> **Related NSF Initiatives**
>
> NSF supports a wide range of biological research, from the subcellular to the biosphere. NSF's Biology Directorate recognizes that "[a]chieving a coherent understanding of the complex biological web of interactions that is life is a major challenge of the future. This challenge will require that knowledge about the structure and dynamics of individual biological units, networks, subsystems, and systems be compiled and connected from the molecular to the global level and across scales of time and space." [a]
>
> NSF initiatives that respond to these challenges include:
> - The *Reintegrating Biology* workshops in 2019 solicited input from the biological research community regarding new research questions that could be addressed by combining approaches and perspectives from different subdisciplines of biology, key challenges and scientific gaps that would need to be addressed to answer these questions, and the physical infrastructure and workforce training needed.[b]
> - The understanding of the *Rules of the Life* program, one of 10 "Big ideas" initiated by NSF in 2016, aims "to identify the casual predictive relationships that could be 'rules of life'. . . [and] to train the next generation of researchers to work across scales and scientific disciplines and to foster convergent research" (NASEM, 2023, p. 4).
> - The ongoing *Biology Integration Institutes* program supports diverse collaborative teams that conduct research, education, and training on critical questions spanning multiple disciplines within and beyond biology.[c] Biology Integration Institutes are designed to integrate across subdisciplines and scales in biology, including spatial, temporal, and biological scales. For example, Advancing Spectral biology in Change ENvironments to understand Diversity (ASCEND) uses spectral information as a common data type (collected from satellites, airborne platforms, uncrewed aerial vehicles and handheld sensors) to decipher the causes and consequences of plant diversity in changing environments within and across individuals, communities, ecosystems, and continents. Enabling Meaningful External Research Growth in Emergent Technologies (EMERGE) examines biological interactions and their responses to change by employing a "'genes-to-ecosystems-to-'genes" approach. Specifically, the team investigates how the genetic and metabolic features of microbial communities, and their biotic and abiotic interactions, give rise to ecosystem outputs, specifically carbon gas emissions, and how ecosystems in turn impact gene expression and allele frequencies. The Regional OneHealth Aerobiome Discovery Network (BROADN) works to understand the microbiome of the air and how it impacts human, animal, and environmental health.
>
> *continued*

> **BOX 1-1 Continued**
>
> - The program Macrosystems Biology and NEON-Enabled Science (MSM-NES): Research on Biological Systems at Regional to Continental Scales, which ran from FY 2010 to FY 2024, supported research on biosphere processes at regional to continental scales and activities to broaden participation of researchers in Macrosystems Biology and NEON-Enabled Science. This program enabled numerous research projects directly linked to continental-scale biology.[d]
>
> ---
>
> [a] See https://www.nsf.gov/bio/about.jsp (accessed November 3, 2023).
> [b] See https://www.nsf.gov/awardsearch/showAward?AWD_ID=1940791&Historical%20Awards=false (accessed October 17, 2023).
> [c] See https://new.nsf.gov/funding/opportunities/biology-integration-institutes-bii (accessed October 17, 2023).
> [d] See https://new.nsf.gov/funding/opportunities/macrosystems-biology-neon-enabled-science-msb-nes/503425 (accessed December 6, 2023).

in a multidisciplinary manner. This report aims to identify the key challenges and the opportunities that could be realized by overcoming these challenges, providing a series of recommendations that will help establish and advance the emerging field of CSB. The report builds on the successes, challenges, and lessons learned from the many networks and synthesis centers across multiple scales that are already in operation.

For CSB to be useful for addressing specific questions and informing decision making, improved methods will be needed to synthesize the vast amounts of data that have been and will be collected. Conceptual frameworks and theories are needed to provide road maps for organizing data to test hypotheses and address central questions. Emerging technologies will need to be deployed. For example, digital twin platforms may create high-fidelity, real-time digital replicas of biological systems that enable research to conduct virtual experiments that are impractical or impossible in the real world (Goodchild et al. 2024). Collaborative networks can provide a platform for scientists to share data, harmonize methodologies, and synthesize existing information to enhance data quality, quantity, and comparability.

Combining perspectives across disciplines is another powerful means to advance understanding. Some of the most rapid progress that has been made is at the intersection of biology and other disciplines, including the work being done in the synthesis centers mentioned above. The National Research Council report *A New Biology for the 21st Century* recommended the "re-integration of the many sub-disciplines of biology, and the integration into biology of physicists, chemists, computer scientists, engineers, and mathematicians to create a research community with the capacity to tackle a broad range of scientific and societal problems" (NRC 2009). The decade and a half since this report's publication has seen movement in this direction in both the conduct and the culture of biology. One example comes from the convergence of ecology and engineering that led to NEON. Further, the development of new fields of study—such as landscape genetics, ecological forecasting, macroevolution, macroecology and macrosystems biology, and engineering biology—and the general rise of open science, open data, and researchers' embrace of large-scale collaborations in the United States and many other countries are enabling the emergence of a multiscale approach to biological research.

Committee's Statement of Task

The committee's statement of task is as follows:

> An ad hoc committee of the National Academies of Sciences, Engineering, and Medicine will conduct a consensus study to identify how biological research at multiple scales can inform the development of a continental scale biology. The committee will convene a series of virtual community workshops to inform its deliberations.
>
> Specifically, the committee will identify and discuss:
>
> - Practices that have been used successfully to translate knowledge, approaches, and tools from small-scale biological research to regional- and continental-scale, and vice versa. For example, how might understanding fine-scale biological discoveries alter specific measurements or experiments that researchers undertake at larger scales and vice versa?
> - Challenges that prevent uptake of these practices.
> - Specific research questions that could serve as pilots for implementing research projects that integrate one or more successful practices.
>
> Finally, the committee will review and refine the practices and questions into a set of recommendations for the research community, funders, and decision makers.

Nominations for the committee were invited through mailing lists maintained by the Board on Environmental Studies and Toxicology and the Board on Life Sciences; from members of those boards and the National Academy of Sciences; professional societies; and National Academies staff. Expertise represented on the committee includes ecology, macrosystems biology, organismal biology, genetics, cell biology, microbiology, biochemistry, molecular biology, data science, computer science, and social science. Committee member biosketches are presented in Appendix A.

COMMITTEE'S APPROACH TO THE STATEMENT OF TASK

The committee approached the statement of task by first identifying key characteristics that define CSB (Box 1-2). As stated in the definition, CSB focuses on biological systems across scales, which are an integral part of coupled human and natural systems; that is, CSB takes a systems approach (Liu et al. 2015).

> **BOX 1-2**
> **Characteristics That Define Continental-Scale Biology**
>
> Continental Scale Biology (CSB) addresses questions about biological processes and patterns that emerge at broad *organizational*, *spatial*, and/or *temporal scales*, that cannot be answered by observations and experiments conducted at either fine or large scales alone. CSB inherently incorporates *multiple scales*, from the subcellular to the global biosphere (Figure 1-1), from the local to global spatial extents, from less than a second to millennia. Specific CSB research may operate across one, two, or all three kinds of scales: organizational, spatial, or temporal.
>
> Further, CSB treats biological systems as part of coupled human and natural systems, given widespread human impacts and intensifying human–nature interactions worldwide.
>
> CSB is enabled by emerging theory; recent developments of experimental and observational networks, tools, and analytical techniques; and changes in the culture of biological science that facilitate collaboration among multidisciplinary teams with members from around the globe.

CSB is an interdisciplinary frontier. An important task is to address the opportunities and challenges associated with the necessary theoretical, empirical, and cultural integration of CSB with allied disciplines. These include natural sciences such as hydrology (Brutsaert 2023); computer engineering (Wright and Wang 2011); and social sciences such as behavioral science (Fischhoff 2021), economics (Barrett 2022, Dasgupta 2021, Polasky et al. 2019), geography (Cutter 2024, Fotheringham 2024), governance (Bebbington et al. 2020, Brondizio et al. 2009), and political science (Agrawal et al. 2023). This integration is essential not only to making scientific advancements in CSB but also to promoting its application to biological, environmental, and societal problems across local, regional, national, continental, and global scales.

To inform these efforts, the committee organized three public information-gathering meetings. The first, on April 24–25, 2023, focused on frontier research efforts demonstrating CSB. The second, on June 15, 2023, reviewed applications of networks, analytical and sampling tools, and data integration approaches to CSB, and challenges that limit their application. The third, on August 21, 2023, reviewed data collection, collective engagement and involvement, and innovative tools and techniques to aid in the advancement and understanding of CSB. Agendas for all three are presented in Appendix B.

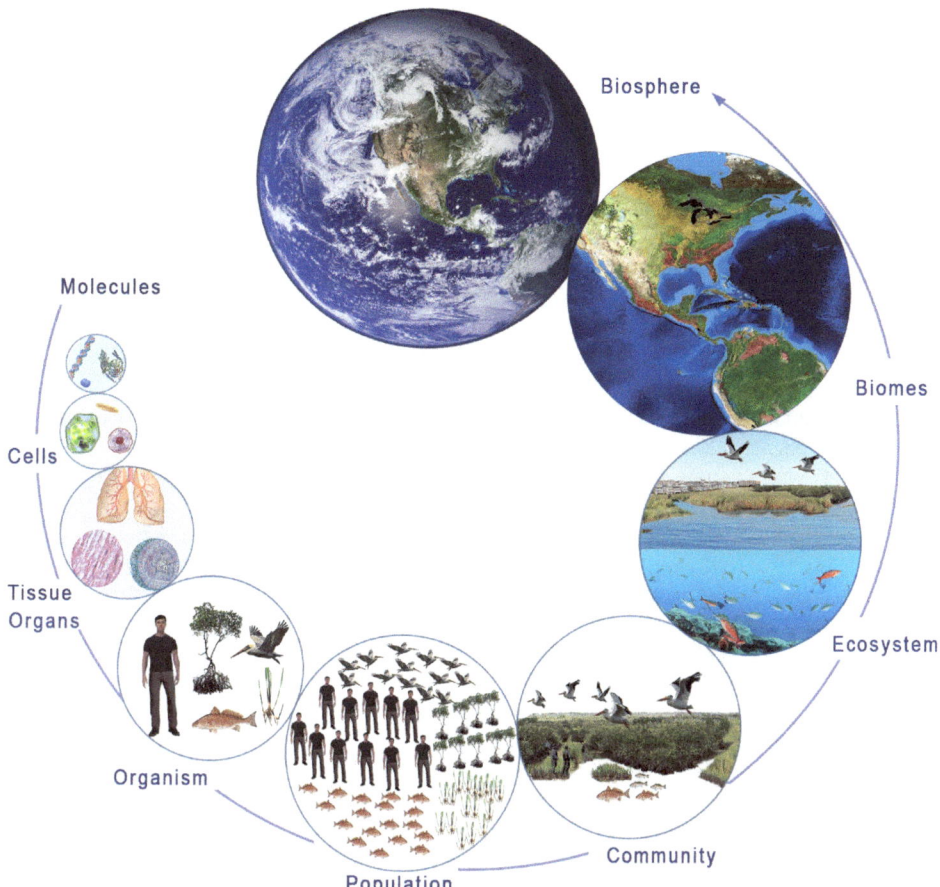

FIGURE 1-1 Biological organizational scales of continental-scale biology.

Report Organization

Chapter 2 presents four themes that represent areas where CSB could make the most impact, with examples of specific research questions that could serve as pilots for implementing research projects that integrate one or more successful practices. This chapter also presents examples of research and management efforts corresponding to and integrating the four themes.

Chapter 3 describes how theory can inform the development of observational and experimental programs for CSB and vice versa, and the development of the resources needed to support those programs.

Chapter 4 evaluates the tools and networks needed to conduct biological research across multiple organizational, spatial, and temporal scales, and identifies new tools or enhancements that could more fully realize the promise of CSB.

Chapter 5 identifies training and capacity-building needs and challenges to support CSB.

Chapter 6 presents overarching recommendations and a vision for realizing CSB.

REFERENCES

Agrawal, A., J. Erbaugh, and N. Pradhan. 2023. The Commons. *Annual Review of Environment and Resources* 48:531-558. https://doi.org/10.1146/annurev-environ-112320-113509.

Barrett, S. 2022. A biodiversity hotspots treaty: The road not taken. *Environmental and Resource Economics* 83:937-954. https://doi.org/10.1007/s10640-022-00670-5.

Bebbington, A., A. Chicchon, N. Cuba, and L. Sauls. 2020. Priorities for governing large-scale infrastructure in the tropics. *Proceedings of the National Academy of Sciences of the United States of America* 117: 21829-21833. https://doi.org/10.1073/pnas.201563611.

Bonan, G.B. 2008. Forests and climate change: Forcings, feedbacks, and the climate benefits of forests. *Science* 320:1444-1449. https://doi.org/10.1126/science.1155121.

Brondizio, E.S., E. Ostrom, and O.R. Young 2009. Connectivity and the governance of multilevel social-ecological systems: The role of social capital. *Annual Review of Environment and Resources* 34:253-278. https://doi.org/10.1146/annurev.environ.020708.100707.

Brutsaert, W. 2023. *Hydrology: An Introduction,* 2nd ed. Cambridge, UK: Cambridge University Press.

Cutter, S. 2024. The origin and diffusion of the Social Vulnerability Index (SoVI). *International Journal of Disaster Risk Reduction* 109:104576. https://doi.org/10.1016/j.ijdrr.2024.104576.

Dasgupta, P. 2021. *The Economics of Biodiversity: The Dasgupta Review.* HM Treasury, London.

Fischhoff, B. 2021. Making behavioral science integral to climate science and action. *Behavioural Public Policy* 5(4):439-453. doi:10.1017/bpp.2020.38.

Fotheringham, A.S. 2024. How to solve the scale "problem" in spatial analytics. Pp. 55-66 in *A Research Agenda for Spatial Analysis,* L.J. Wolf, R. Harris, and A. Heppenstall, eds. Cheltenham. UK: Edward Elgar.

Goodchild, M. F., D. Connor, A. S. Fotheringham, A. Frazier, P. Kedron, W. Li, and D. Tong. 2024. Digital twins in urban informatics. *Urban Informatics* 3:16. https://doi.org/10.1007/s44212-024-00048-6.

Heffernan, J.B., P.A. Soranno, M.J. Angilletta, Jr., L.B. Buckley, D.S. Gruner, T.H. Keitt, J.R. Kellner, J.S. Kominoski, A.V. Rocha, J. Xiao, T.K. Harms, S.J. Goring, L.E. Koenig, W.H. McDowell, H. Powell, A.D. Richardson, C.A. Stow, R. Vargas, and K.C. Weathers. 2014. Macrosystems ecology: Understanding ecological patterns and processes at continental scales. *Frontiers in Ecology and the Environment* 12:5-14. https://doi.org/10.1890/130017.

LaRue, E., J. Rohr, J. Knott, W. Dodds, K. Dahlin, J. Thorp, J. Johnson, M. Rodriguez-Gonzalez, B. Hardiman, M. Keller, R. Fahey, J. Atkins, F. Tromboni, M. SanClements, G. Parker, J. Liu, and S. Fei. 2021. The evolution of macrosystems biology. *Frontiers in Ecology and the Environment* 19:11-19. https://doi.org/10.1002/fee.2288.

Liu, J., H. Mooney, V. Hull, S.J. Davis, J. Gaskell, T. Hertel, J. Lubchenco, K.C. Seto, P. Gleick, C. Kremen, and S. Li. 2015. Systems integration for global sustainability. *Science* 347:1258832. https://doi.org/10.1126/science.1258832.

NASEM (National Academies of Sciences, Engineering, and Medicine). 2023. Understanding the Rules of Life Program: Scientific Advancements and Future Opportunities. Washington, DC: The National Academies Press. https://doi.org/10.17226/27021.

NRC (National Research Council). 2009. *A New Biology for the 21st Century*. Washington, DC: The National Academies Press.

Polasky, S., C.L. Kling, S.A. Levin, S.R. Carpenter, G.C. Daily, P.R. Ehrlich, G.M. Heal, and J. Lubchenco. 2019. Role of economics in analyzing the environment and sustainable development. *Proceedings of the National Academy of Sciences of the United States of America* 116:5233-5238. https://doi.org/10.1073/pnas.190161611.

Stark, S.C., D.D. Breshears, E.S. Garcia, D.J. Law, D.M. Minor, S.R. Saleska, A.L.S. Swann, J. C. Villegas, et al. 2016. Toward accounting for ecoclimate teleconnections: Intra- and inter-continental consequences of altered energy balance after vegetation change. *Landscape Ecology* 31:181-194. https://doi.org/10.1007/s10980-015-0282-5.

Wright, D.J., and S. Wang. 2011. The emergence of spatial cyberinfrastructure. *Proceedings of the National Academy of Sciences of the United States of America* 108:5488-5491. https://doi.org/10.1073/pnas.110305110.

2

Themes for a Continental-Scale Biology

The committee identified four major themes in the types of research that CSB is particularly well positioned to support and in each theme, identified a series of research questions that could serve as pilots for implementing research projects. These themes and questions are neither mutually exclusive nor comprehensive, but rather are intended to stimulate discussion about directions for the development of the field (Figure 2-1). Although these themes have been extensively studied, previous research tended to focus on one particular scale. Thus, comparisons and interactions among different scales have not received sufficient attention. Furthermore, the literature on these themes is largely from biological and ecological perspectives. CSB aims to systematically and explicitly expand from a single scale to multiple scales, and from biological and ecological perspectives to integrated biological, ecological, and socioeconomic perspectives to account for pervasive human–nature interactions.

The four major research themes are interrelated, as shown in Figure 2-1. For instance, the fourth theme, "'sustainability of ecosystem services'" is related to the first three because biodiversity and ecosystem functions are essential to produce ecosystem services, resilience and vulnerability are key in maintaining ecosystem services, and connectivity is central to shaping distribution and use of ecosystem services across space and over time.

Understanding scalability and how biological properties (e.g., patterns and processes) vary or remain the same across scales is an important area of inquiry for CSB. While different themes may have specific questions, a common question for all themes is how different biological properties vary or do not vary across spatial, organizational, and temporal scales. Furthermore, addressing all themes can benefit from a systems approach.

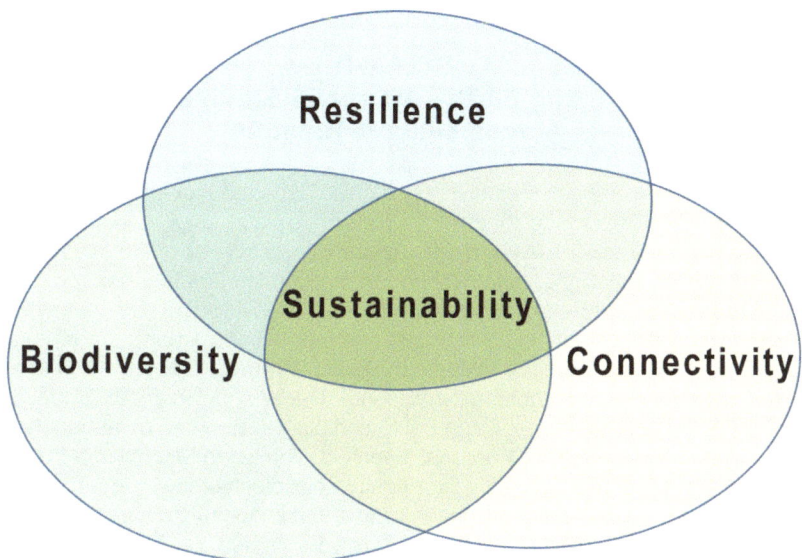

FIGURE 2-1 Relationships among the four themes. Sustainability lies at the center, incorporating elements of biodiversity, resilience, and connectivity. The latter three themes overlap with each other while also having their own domains of inquiry.
SOURCE: Stacy Jannis.

BIODIVERSITY AND ECOSYSTEM FUNCTION

A pivotal theme of ecological research over the past half century has been the relationship between biological variation, or biodiversity (in terms of taxonomy, function, phenotype, genotype or phylogenetic placement) and ecosystem functions, for example, the cycling of carbon, nitrogen, phosphorus, sulfur, oxygen, water, and trophic interactions.

Much of the experimental work has been conducted at local scales, but generalities have emerged to indicate that biodiversity is important to ecosystem function at multiple scales. As yet, there is scant evidence available to define these relationships and to decipher how they may change across scales—information that will be central to determining how biodiversity across spatial and temporal scales drives local to Earth system functions. An example comes from recent work to catalog Earth's microbial biodiversity to better understand the function and structure of microbial communities that influence processes across entire ecosystems, from the gut microbes that help digest food to the microbes that help cycle carbon and nitrogen (see Box 2-1).

Ecosystem studies have tended to focus on collocated measures of biodiversity and ecosystem processes, such as the Jena experiment in Germany[1] and the Cedar Creek experiment in the United States.[2] CSB goes beyond typical ecosystem studies

---
[1] See https://the-jena-experiment.de (accessed August 16, 2024).
[2] See https://cedarcreek.umn.edu/about-cedar-creek-lter (accessed August 16, 2024).

> **BOX 2-1**
> **Cataloging Earth's Microbial Biodiversity**
> **Across Multiple Habitats**
>
> The majority of Earth's biodiversity is microbial, and increasingly, scientists are realizing that microbes (bacteria, archaea, protists, fungi, and viruses) are the drivers of key processes in plants, animals, and across entire ecosystems, from the gut microbes that help digest food to the microbes that help cycle carbon and nitrogen. Understanding these processes will require new knowledge of the organization, evolution, functions, and interactions among the billions of the planet's microbial species. Yet most microbes remain unknown to humans: a 2016 paper in *Proceedings of the National Academy of Sciences* (Locey and Lennon 2016) estimated that Earth could contain 1 trillion microbial genotypes, with only one-thousandth of 1 percent identified. The Earth Microbiome Project,[a] founded in 2010, seeks to catalog Earth's microbial biodiversity by identifying and classifying microbes and identifying the factors that shape microbial communities.
>
> The project leverages the power of crowd sourcing, relying on thousands of researchers around the world to collect samples from organisms and environments as diverse as the human gut, a bird's mouth, the soil of an Antarctic valley, a lake in Germany, and the bottom of the Pacific Ocean. In a 2017 *Nature* publication, Thompson et al. (2017) detailed protocols, analytical methods, and software developed to process the 200,000 samples collected. The team decided to focus on bacteria, sequencing a 90-base-pair region of the 16S ribosomal RNA gene to act as a unique identifier of each species present in the sample.
>
> The data generated have yielded new insights about the organizing principles that govern the structure of microbial communities. For example, Thompson et al. (2017) compared samples taken from different environments, finding that salinity and host association (e.g., the microbes were found in plants and animals) play significant roles as environmental factors that differentiate microbial communities.
>
> In a 2022 *Nature Microbiology* publication, the Earth Microbiome Project team expanded the analysis to identify the metabolites present in each sample (Shaffer et al. 2022). Microbes produce metabolites as a byproduct of the reactions they are carrying out. By identifying and quantifying the different metabolites in each sample, researchers can learn about microbes' interactions and activities. Working with an additional 880 samples taken from sites around the world, the team used techniques including gas chromatography-mass spectrometry to study the metabolite profile of each sample. In addition to sequencing a region of the 16S ribosomal RNA gene, in this study, the team also sequenced regions of the 18S ribosomal RNA and internal transcribed spacer genes to serve as an identifier for protist and fungal species.
>
> The team developed standardized protocols for each technique, which will make it easier for researchers to compare data from different studies to provide a global-scale view of microbial communities across multiple habitats. All the data generated by the Earth Microbiome Project are available on an open-source platform, and teams of researchers around the world continue to gain new insights from the vast dataset.
>
> ---
> [a] See https://earthmicrobiome.org/ (accessed March 8, 2024).

by integrating collocated measures in the focal place, such as an experiment site, and measures of flows (of organisms, materials, information, water, etc.) between the focal place and other places (e.g., integration of movement ecology with ecosystem ecology). This integration requires new study designs that account for measures of flows and reciprocal effects of flows and internal factors as indicated in collocated measures.

A CSB approach offers the opportunity to use emerging tools, including satellite remote sensing, genetic sequencing, multi-omics, artificial intelligence, and automated monitoring systems, together with developing theory, to examine the linkages between biodiversity and ecosystem functions across scales, and to learn how these relationships may shift in the face of global change. The committee identified the following example research questions in this theme:

- How does complexity of a biological system at one scale influence emergent properties of the system at the next scale, and how do those properties feed back to influence complexity and variation at all scales?
- How do ecosystem functions emerge from biological complexity, what mechanisms are involved, and how do these vary across temporal and spatial scales? How do major global changes, including climate change, human-induced land- and water-use changes and the spread of pests and pathogens, impact how the diversity of life influences and interacts with functions and processes of emergent systems?
- How do human-induced changes in land use and water use, habitat fragmentation and, conversely, habitat connectivity at local and regional scales affect continental-scale biodiversity?
- How does theory support our understanding of the causes and consequences of continental-scale biodiversity in the context of major global changes and human activities?
- How does the distribution of composition, diversity, and complexity of ecosystems influence the cycling of carbon, nutrients, and water at continental scales, and how do these cycles feed back to influence the composition, diversity, and complexity of local systems?
- How do internal factors in a focal place and flows between focal and other places influence biodiversity and ecosystem functions across scales?

RESILIENCE AND VULNERABILITY

Resilience is the capacity of a system to withstand or recover from human and environmental disturbances[3] (Pickett et al. 2013, Reid et al. 2014). Examples include climate change, pollution, habitat loss, and landscape fragmentation, which are impacting ecosystems all over the planet (Mann et al. 2008, Melillo et al. 2014). The effects of human and environmental disturbances can be identified as "press," which means they are chronic, or "pulse," which means they are acute (Inamine et al. 2022). Resilience to

---

[3] See https://www.resalliance.org/resilience (accessed September 3, 2024); see also Levin (2024).

disturbances consists of three components: stability, resistance, and recovery. A stable ecosystem can resist disturbance by maintaining structure, function, and composition. A resilient ecosystem can recover from disturbances over time and essentially return to its original state. Ideally, multiple metrics should be used to provide a composite assessment (Dakos and Kéfi 2022). In contrast, vulnerability is the extent to which a system (or a system component) is likely to experience harm due to exposure to disturbances (Turner et al. 2003). A vulnerable system is sensitive to disturbances and lacks adaptive capacity when conditions change.

Much research on resilience and vulnerability in biological systems has been conducted at a particular scale. However, little evidence exists regarding how resilience, vulnerability, and their relationships may change across scales. Such information is critical to understanding sustainability and many other aspects of biological systems across different scales. A CSB approach offers the opportunity to study resilience, vulnerability, and their relationships across scales, and to examine how these relationships may vary under global change, human activities, and connectivity. CSB research can also be used to assess the sensitivity of biological systems and their adaptive capacity to resist, recover, or change in biological composition and function across multiple scales in response to disturbances.

Resilience and vulnerability are affected by factors not only within a place but also in adjacent and distant places. For example, population dynamics in a place are affected by not only internal factors such as birth rates but also interactions with adjacent and distant places such as migration. If the population in the focal place is eliminated or reduced by a disturbance, the population in the focal place may or may not recover without immigrants from adjacent and distant places. Such immigrants may be affected by human activities, such as human-assisted migration of species in response to the disturbance.

Understanding the factors that influence the resilience and vulnerability of individuals, populations, communities, and ecosystems across scales to changing conditions will be essential for informing multiscale strategies for the conservation of biodiversity and management of ecosystem services. One example comes from the research that explores how the microbial communities that make up coral reef biofilms could provide an indicator of the health and resilience of the reef ecosystem (see Box 2-2). Another describes the effects of environmental and human factors on vulnerability of native plant populations to invasive species (Box 2-3).

Critical questions remain to assess a system's resilience and vulnerability that through drivers, processes or impacts are fundamentally linked to CSB:

- How do biodiversity and human activities influence resilience of biological systems at multiple spatial and temporal scales? How does this relationship shift across scales of biological organization, from the population (diversity within a species) to the biome?
- Which ecosystems and functions at what scales are most exposed and vulnerable to natural and human disturbances?
- What are the tipping points and thresholds of resilience and vulnerability across scales? What continental-scale drivers affect the spatial and temporal patterns of

ecosystem vulnerability at local, regional, and national scales, and how do local to regional drivers affect vulnerability and resilience at the continental scale?
- How does the resilience or vulnerability of one system—for example, to land use or climate change—affect resilience or vulnerability of other systems near and far?

CONNECTIVITY

Ecosystems do not exist in isolation. The world is more connected due to a variety of factors such as globalization, climate change, land and water use, disease spread, and environmental changes. Such connections have important biological and socioeconomic consequences across multiple scales, including the continental scale. Increasingly, scientists are understanding that human–nature interactions or natural events in one place can have profound effects on ecosystems in other places, both nearby and distant. For example, deforestation may be facilitated by adjacent roads (Cropper et al. 2001) and distant demand for agricultural products (DeFries et al. 2010). Wildfires have increased in many places (MacDonald et al. 2023) and can influence air quality in areas far from the actual fires (Jaffe et al. 2020); forest wildfires in Oregon have affected air quality on the East Coast of the United States (Miller, 2021), and the Canadian wildfires in 2023 caused air quality alerts as far south as the Carolinas (Dennis and Koh, 2023).

Many studies have examined linkages in time and space. Examples include the stream continuum concept (Stallard 1998), transoceanic transport of nutrients (Vitousek 2004), and seabird movement of pelagic nutrients into land and intertidal rocky shores (Healing et al. 2024). However, more systematic approaches are needed to effectively integrate biota, habitats, ecosystem functions and materials, and humans across time and space. For example, while spatial connections in metapopulations (and metacommunities and metaecosystems) have been extensively studied (Holt 1997, Leibold et al. 2004, Loreau et al. 2003), their main focus has been on ecological dimensions with little explicit attention to human dimensions. Furthermore, previous studies usually do not provide an explicit way of contrasting spatial linkages (e.g., adjacent and distant). The explicit integration can help scientists understand, avoid, and solve problems when, for example, sustainability efforts focus on one ecosystem issue without realizing how that may impact connected systems near and far.

CSB can offer tools that can combine data and research from various realms across time and space, helping to build a stronger understanding of the system as a whole. For example, interdisciplinary frameworks provide systematic ways to combine data and insights from different ecosystems across time and space. A number of frameworks have been developed to address multiscale and cross-scale issues (e.g., Allen and Starr 1982, Folke et al. 2011, Heffernan et al. 2014, Levin 1992, Peters et al. 2008, Rose et al. 2016). They have provided useful insights and inspired research efforts on CSB. However, they do not explicitly specify human–nature interactions and feedbacks within as well as between adjacent and distant systems, which are widely recognized common real-world phenomena at multiple scales.

BOX 2-2
Biodiversity in Coral's Slimy Biofilms as an Indicator of Reef Resilience

Coral reefs harbor millions of bacteria, both floating free in the water and forming a slimy reef coating called biofilm. Understanding the mix of microbes that make up coral biofilm could provide crucial insights on the ecosystem services they provide—and according to recent research, could also provide a measure of the resilience of the reef ecosystem (Remple et al. 2021).

Biofilm is a bacterial matrix that coats surfaces—in this case, coral reefs. The microbes that make up biofilm keep coral reefs healthy: biofilm bacteria help with organic matter decomposition, nutrient cycling, and recruiting the next generations of coral larvae to sustain reef structure. These ecosystem services are likely governed by the composition of the biofilm, a delicate balance of different species of bacteria that depends on environmental conditions, including factors such as fluid dynamics and nutrient availability.

Until recently, few studies have focused on the factors that influence coral biofilm composition. More is known about free-floating marine microbes. For instance, research has shown that ocean acidification and warming can cause a decrease in microbial diversity in some coral ecosystems, and other studies have shown that nutrient pollution can boost populations of disease-causing bacteria.

In combination with other stressors such as overfishing, nutrient pollution also can boost the growth of fleshy seaweeds called macroalgae in coastal ecosystems. Once the macroalgae starts to grow, it can launch a feedback loop that shifts the ecosystem from coral dominated reefs to those dominated by algae. The exact mechanisms of this are still being worked out, but differences in the metabolic processes carried out by macroalgae and corals lead to differences in the composition of the organic matter they exude into the water column. Macroalgae in the ecosystem change the physical and chemical qualities of the water, which restructures microbial communities and favors a less diverse mix of microbial species that is less able to support coral vitality.

Remple et al. (2021) investigated how nutrient pollution and algae impact the composition of coral biofilm. The team cultured biofilms in environments containing sand, algae, or corals and with different levels of added nutrients, and surveyed the bacterial communities in each biofilm sample.

The study found that at earlier stages, biofilms were dominated by "metabolic generalists"—microbes that are less specialized in the use of organic carbon compounds. Over time, biofilm communities became more diverse and contained markedly different microbial communities than the free-floating bacteria within the same samples. Adding nutrients decreased the diversity of biofilm communities, favoring primary producers and bacteria that thrive in nutrient-rich conditions rather than the nutrient-cycling bacteria typically found in biofilms.

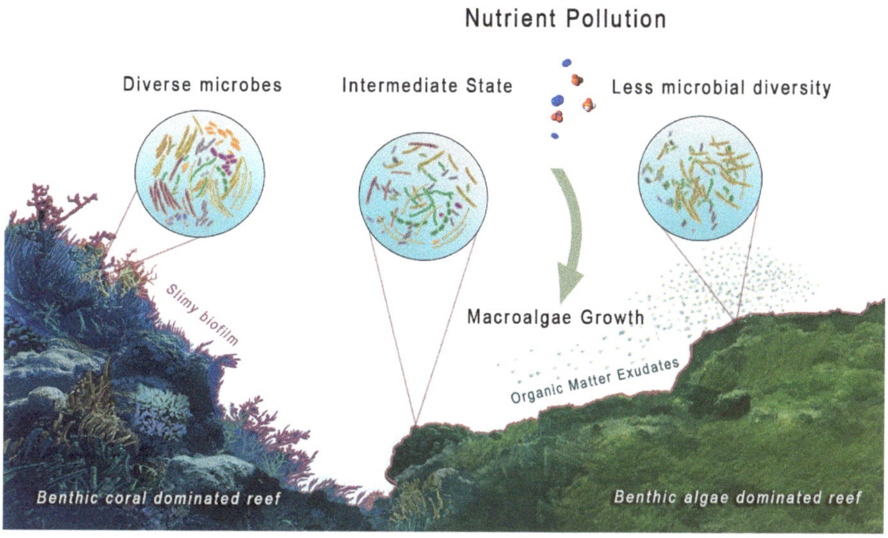

FIGURE 2-2-1 Diverse microbial communities make up coral reef biofilms. They keep coral reefs healthy by helping with organic matter decomposition, nutrient cycling, and recruiting the next generations of coral larvae to sustain reef structure. However, nutrient pollution and the presence of fleshy, seaweed-like macroalgae in the reef ecosystem can lead to a decrease in biofilm microbial diversity—and in the resilience of the corals.
SOURCE: Stacy Jannis.

Biofilms cultured with algae were less diverse than those grown with corals, and were dominated by microbial species adept at growing quickly on nutrient-rich carbon compounds. In contrast, the biofilms cultured with corals maintained a rich diversity of microbial life throughout the experiment, although the addition of nutrients caused a decrease in diversity in mature biofilms. This indicates that the addition of nutrients could derail microbial diversity even in coral-dominated reefs.

Finding that distinct differences exist between the composition of biofilms in coral-dominated reefs and those that include algal species paves the way for future understanding of how shifts in coral ecosystems may impact key microbial processes that are crucial for reef resilience.

Biofilm biodiversity could also potentially provide an indicator of ecosystem health: a more resilient ecosystem may be able to maintain biofilm diversity even as environmental conditions change, and conversely, a loss of biofilm diversity could indicate that environmental conditions are decreasing ecosystem function.

> **BOX 2-3**
> **Effects of Local, Landscape, and Regional
> Drivers of Vulnerability to Plant Invasions**
>
> Leveraging multiscale data on native and non-native plant species across National Ecological Observatory Network sites, Ibáñez et al. (2023) quantified the compound effects of local, landscape, and regional drivers of vulnerability to plant invasions. Vulnerability was influenced by temperature, precipitation, productivity, and human modification, but the magnitude and nature of these influences varied widely across ecoregions. For example, in colder regions, native species were more vulnerable to local-scale losses, probably due to shorter growing seasons favoring invasive plants, whereas in warmer regions, the impact of non-natives was more pronounced at landscape scales, suggesting that local non-native effects can extend over larger areas. By accounting for cross-scale effects, the study provided a nuanced view of vulnerability that could aid in localized and broader-scale decision making for managing plant invasions. The results also underscore the importance of considering multiple scales and environmental factors to better predict and manage the impacts of invasive species on native plant communities.

A recent framework that built upon, expanded from, and integrates various previous frameworks is the metacoupling framework (Liu 2023), which explicitly considers the interactions that take place within a system (including human–nature and nature–nature interactions) and interactions with adjacent and distant systems. Nature includes both biotic components (animals, plants, and microorganisms) and abiotic components (e.g., air, climate, geophysical conditions, soil/land, water). Humans consist of social, economic, political, cultural, and other components. The framework takes a systems approach (e.g., by treating each place as a coupled human and natural system, and all interacting places near and far as metacoupled human and natural systems). It integrates social, economic, and ecological connections and feedback within and between systems and provides a way to enhance synergies and reduce trade-offs between them. This framework has been applied at multiple scales ranging from local and regional scales (Wang et al. 2024, Zhao et al. 2021) to national and global scales (Carlson et al. 2020, Xiao et al. 2024, Xu et al. 2020). It is discussed in more detail in Box 2-4, which illustrates how the metacoupling framework can help scientists and policymakers integrate data on the connections between humans and natural ecosystems, and as an example, to inform fisheries management decisions.

As mentioned above, biodiversity, ecosystem functions, resilience, and vulnerability are affected by not only internal factors within a focal place but also factors in adjacent and distant places. The metacoupling framework accounts for all these factors.

It goes beyond the biological or ecological approaches because the movement of species and materials, for example, across boundaries is not just driven by natural factors, but also driven by human factors and human–nature interactions. For example, the materials that move across the landscapes may be released by human activities such as agricultural practices including fertilizer applications. Increasingly, migration of many species is affected by humans.

Currently, the metacoupling approach is limited by a lack of understanding of the rate and quantity of connections and feedback between systems that influence these dynamics. To truly deliver on the promise of metacoupling analysis requires both novel and robust methods of data integration and development and refinement of theory to support such advancements. Research questions to support these goals include:

- How do human–nature interactions vary across organizational, spatial, and temporal scales, and what are the impacts of economic globalization and environmental changes on these dynamics?
- How do human activities and environmental changes (including climate change) in one system generate cascading effects across adjacent and distant systems?
- How do interactions within a focal system, between adjacent systems, and between distant systems create feedback, synergistic, or trade-off effects on resilience and sustainability of biodiversity and ecosystems across scales?
- How do risks associated with invasive species, pathogen emergence, and other shocks such as natural disasters spread across adjacent and distant systems at multiple scales?

SUSTAINABILITY OF ECOSYSTEM SERVICES

Ecosystem services are the benefits people obtain from ecosystems, including the provisioning of food and water; the regulation of floods, drought, land degradation, and disease; nutrient cycling and soil formation; and nonmaterial benefits such as recreational, spiritual, and religious benefits of nature (Costanza et al. 1997, Daily and Matson 2008, Diaz et al. 2019).

Sustainability relates to an ecosystem's ability to continue to provide services, without a substantial reduction in the level of ecosystem services, in the face of environmental and human disturbances or state changes such as land changes due to urbanization (Grimm et al. 2008, Seto et al. 2012, Shepherd et al. 2002, Turner et al. 2007). Being able to assess the sustainability of an ecosystem service across scales is critical for developing targeted multiscale mitigation and adaptation strategies for sustainability. However, previous research and management tended to largely focus on sustainability of ecosystem services at a single scale (e.g., Ashrafi et al. 2022) although there is a strong need for pursuing multiscale management and sustainability (e.g., Clark and Harley 2020, Hou et al. 2023, Tavárez et al. 2022).

CSB can contribute to our understanding of sustainability by examining the processes through which natural systems give rise to ecosystem services across scales, and in turn contribute to inclusive wealth. Inclusive wealth encompasses all the assets

BOX 2-4
A Metacoupling Framework to Help Optimize Salmonid Research, Management, and Policymaking

Salmonids—fish in the salmon family including salmon, trout, char, and freshwater whitefish—are not only key species in their ecosystems, playing roles as both predators and prey, in nutrient cycling, and as indicators of ecosystem impairment, but are also economically and culturally important as a source of food for people and a target for recreational fishing. Increasingly, anthropogenic factors such as climate change, groundwater withdrawal, and habitat disruption are impacting salmonid populations.

Understanding how these factors interact is important for monitoring the health of salmonid fisheries, but to date most fisheries research does not have a systematic way to integrate the influences of biota, habitats, and humans across time and space. Instead, most sustainability efforts have solved problems one at a time, often at the cost of other components of the system.

Recent advances in theory and methods have helped develop a "metacoupling" approach to studies of coupled human and natural systems (CHANS) that could offer insights for salmonid management and policymaking. The metacoupling framework would provide a useful tool for studying the interactions between humans and nature that occur within a place (in this example, the salmonid fisheries), and between adjacent places and distant places, allowing the analyses to scale from local to global levels.

Overall, the metacoupling framework seeks to integrate social and ecological information, characterize trade-offs and feedbacks across scales, and understand the roles that different members of the community play, including anglers, landowners, and people who work in the recreational fishing industry. The ultimate goal is a holistic, "big picture" view that informs fishery management decisions to improve the relationships between human and natural systems as a whole, rather than in specific physical places in isolation, to better sustain salmonid fisheries locally, regionally, and globally.

This metacoupling approach was illustrated by Carlson et al. (2020), which analyzed brook char and brown trout fisheries in two adjacent cold-water streams in Michigan, the Twin and Chippewa creeks. These salmonid species have enormous socioeconomic and recreational significance in the region, representing a crucial component of statewide fisheries that generate $4.2 billion in overall economic effect. At the same time, there are concerns that groundwater withdrawal to supply a drinking water bottling plant could alter stream hydrology, which has caused controversy among landowners and anglers over how the hydrologic changes may impact salmonid populations.

Using the metacoupling framework, the study authors combined information on how water, information, fish, people, and money are moving through individual CHANS (such as Twin Creek), between adjacent CHANS (such as Twin and Chippewa creeks), and, at the regional level, through non-adjacent CHANS (such as Twin Creek and Hoffman Creek).

One insight from this analysis is that although statewide regulation on water withdrawals is effective, local community members strongly prefer local-scale governance because it engages local communities and considers the groundwater dynamics of individual streams and watersheds.

Another insight was about a proposed transition from riparian forest to shrubs on the banks of the Twin and Chippewa Creeks. This transition could decrease water temperatures by 0.09°C, helping to maintain the cold-water habitat needed for the brown trout and brook char. However, the metacoupling analysis indicated that this land-use change would reduce angling, and the associated revenue streams for local communities, because the natural beauty of the forested streams is highly valued by salmonid anglers and is a key factor in decisions about which stream to visit, underscoring that what may be ecologically beneficial for the fishery may be socially and economically detrimental, and vice versa. This perspective could help fishery managers to identify a holistic solution that balances social, economic, and ecological trade-offs for sustainable salmonid management.

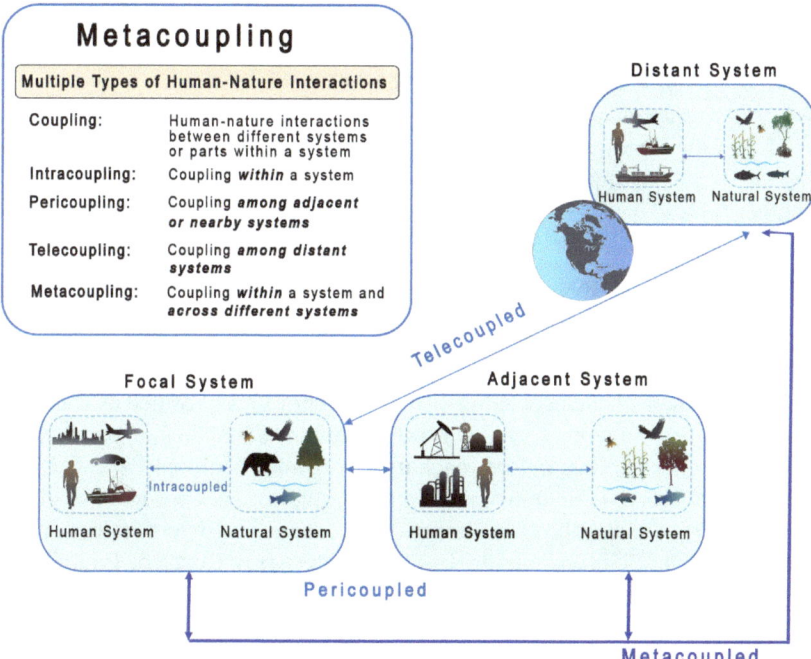

FIGURE 2-4-1 A metacoupling framework provides a tool for studying the interactions between humans and nature. These interactions occur within a focal system such as study sites of the National Ecological Observatory Network (NEON) or Long Term Ecological Research (LTER) sites (intracoupling) and between adjacent systems (pericoupling) and distant systems (telecoupling), allowing the analyses to scale from local to global levels. The different coupling terms provide a clear way to contrast different scales of linkages and feedback.
SOURCE: Stacy Jannis.

that contribute to human well-being—including the natural capital, human capital, and produced capital—and is a measure designed to address whether society is on a sustainable trajectory (Polasky et al. 2015, UNEP 2018). Sustainability can then be defined as nondeclining human well-being, based on metrics of inclusive wealth.

CSB research brings knowledge, via systems' resilience and vulnerability to stressors, on how sustainability of ecosystem services may be affected across organizational, temporal, and spatial scales. For example, trees offer crucial ecosystem services, such as carbon sequestration, air pollution removal, and wood production. Such services vary among different tree species and lineages and different regions of the contiguous United States (Cavender-Bares et al. 2022). Box 2-5 discusses how the ecosystem service of slope stability changes across spatial scales with changing rainfall and species composition, and Box 2-6 describes how tree fecundity affects the long-term maintenance of ecosystem services provided by forested areas.

The committee identified a series of research questions to explore how CSB can increase understanding of sustainability, including:

- Which ecosystem services are most sustainable at different scales in the face of human activities and environmental change?
- How do socioeconomic, urbanization, and land-use change affect ecosystem services across local to continental scales?
- What factors across scales affect ecosystem services at the scale of interest, and how does the sustainability of an ecosystem service shift across spatial scales?
- How does sustainability of ecosystem services at one scale affect sustainability at other scales?

EXAMPLES OF INTERWORKING OF THE FOUR THEMES

Everglades Restoration

Florida's Everglades have been described as a river of grass, formed as waters from Lake Okeechobee flowed slowly southward over sawgrass marshes, cypress swamps, wet prairies, and other habitats before reaching Everglades National Park and Florida Bay. The water that created the unique ecosystem, however, also posed a flooding risk for people. In 1928, the deadly Okeechobee Hurricane sent 15-foot waves from the lake, killing more than 2,500 people. In 1947, the U.S. Army Corps of Engineers (the Corps) proposed the Central and South Florida Project to provide flood protection for growing urban development and agriculture activities in South Florida. By 1960, the Corps had built 720 miles of canals and 1,000 miles of levees that profoundly altered the region's wetlands. Today, at 1.5 million acres,[4] the Everglades is half its original size and is impaired by contaminated runoff from cities and farms (NASEM 2023).

Since the early 1990s, a coalition of local, state, and federal agencies, nongovernmental organizations, local tribes, and citizens has been working to reverse the

---

[4] See https://www.nps.gov/ever/planyourvisit/basicinfo.htm (accessed August 30, 2024).

> **BOX 2-5**
> **Bridging Spatial Scales: How the Ecosystem Service of Slope Stability Changes with Rainfall Amounts and Forest Species Composition**
>
> Bridging spatial scales, from belowground roots to hillslopes, has allowed research to inform how the ecosystem service of slope stability changes with rainfall amounts and forest species composition. Roots reinforce soils against shallow landslide initiation by increasing apparent cohesion. The magnitude of the reinforcement provided by roots depends on the number, the size distribution, and the elastic properties of roots that cross potential failure planes. In forested landscapes, the variability in belowground properties at a hillslope scale makes it challenging to predict root reinforcement at the landscape scale. Topography can affect root strength through changes in root cellulose content, with stronger roots on divergent landscape positions and weaker roots on convergent landscape positions (Hales et al. 2009). Using an ecohydrologic model fused with a landslide model allowed spatial scaling to generate factor-of-safety maps that predict landslide potentials (Band et al. 2012). Further scaling relationships between tree height, root biomass, and dynamic soil moisture increased the model's predictive ability (Hwang et al. 2015). Ultimate model refinements incorporated a demonstrated feed-forward response of root strength to soil moisture; as roots gain moisture content, they become weaker (Hales and Miniat 2017). Landslide maps for counties in western North Carolina were used to develop steep-slope ordinances such that development could not occur on steep slopes with a high likelihood of failure. This work has also been influential in coordinating with the National Weather Service to establish guidelines for when they include wording on landslide hazards in their public advisories for floods and flash floods in the region.

damage to the Everglades. Led by the U.S. Army Corps of Engineers (USACE), and the South Florida Water Management District, the blueprint for the restoration effort, the Comprehensive Everglades Restoration Plan (CERP), was published in 1999. The plan proposed 68 individual projects to be constructed over an estimated 30 to 40 years in the South Florida Ecosystem. CERP's primary goal is to "get the water right"—that is, to deliver the right amount and quality of water to restore characteristics of the historic ecosystem (NASEM 2023). The plans included the creation of below- and aboveground water storage, water quality treatment, and the removal of barriers to the historic sheet flows of water (Figure 2-2).

CERP is a good illustration of the interplay of the four themes laid out in this report, from local to ecosystem scales. At the center is the desire to restore and protect long-term sustainability of the South Florida ecosystem and the services it provides. Wetland ecosystems filter pollutants, excess nutrients, and sediments from the water,

> **BOX 2-6**
> **Effect of Tree Fecundity on the Long-Term Maintenance of Ecosystem**
> **Services Provided by Forested Areas**
>
> The long-term maintenance of ecosystem services provided by forested areas will be highly dependent on the functionality of these systems, that is, the maintenance of healthy ecosystems (Turner and Daily 2008). Tree reproduction rates that keep pace with current environmental changes will be critical for the preservation of these ecosystems, either by maintaining populations in place or by tracking climate change via migration (Davis and Shaw 2001, Martinez-Vilalta and Lloret, 2016). Linking locally collected records of tree fecundity across North America, Clark et al. (2021) built a continental Masting Inference and Forecasting (MASTIF) network of primary data to quantify tree species fecundity across North America. They describe a biogeographic divide between the East, increasing fecundity, and the West, decreasing fecundity. These continental-scale forest responses are driven by both the direct and indirect effects of climate change on individuals via stand-level features. This analysis also shows how, despite drier summers, hot spots of tree fecundity resulted from the interaction between regional climatic trends, spring rising temperatures and declining moisture deficits, and optimal local stand structure that affected individual tree growth. The potential effects of changing environmental conditions on tree fecundity were the result of drivers acting and interacting across spatial scales.

which helps maintain water quality. The wetlands also replenish aquifers that serve the drinking water needs for one-third of Floridians and irrigation for much of the state's agriculture. In addition, the national parks and wildlife refuges offer opportunities for tourism and enhancing health and well-being that people obtain from enjoying nature.

Achieving sustainability in the Everglades requires a high degree of integration of scientific knowledge about the biodiversity of the system. The Everglades are home to more than 360 bird species, 60 reptile species, and 40 mammal species, and a wide variety of plant life, including bromeliads, cacti, succulents, native grasses, seagrasses, and wildflowers. Longstanding CERP challenges include balancing the right quantities of water and timing of flows to habitats hosting the endangered Cape Sable seaside sparrow, providing food and nesting space for the endangered snail kite and wading bird species, and addressing degradation of tree islands and ridge and slough topography by water levels that were too high in some places and too low in others.

A big part of restoration decision making involves consideration of connectivity in the system, including trade-offs, interactions, and synergies across multiple scales in the system from organisms to habitats, to the whole ecosystem and adjacent ecosystems. CERP hydrological and ecological modeling and monitoring efforts strive to examine

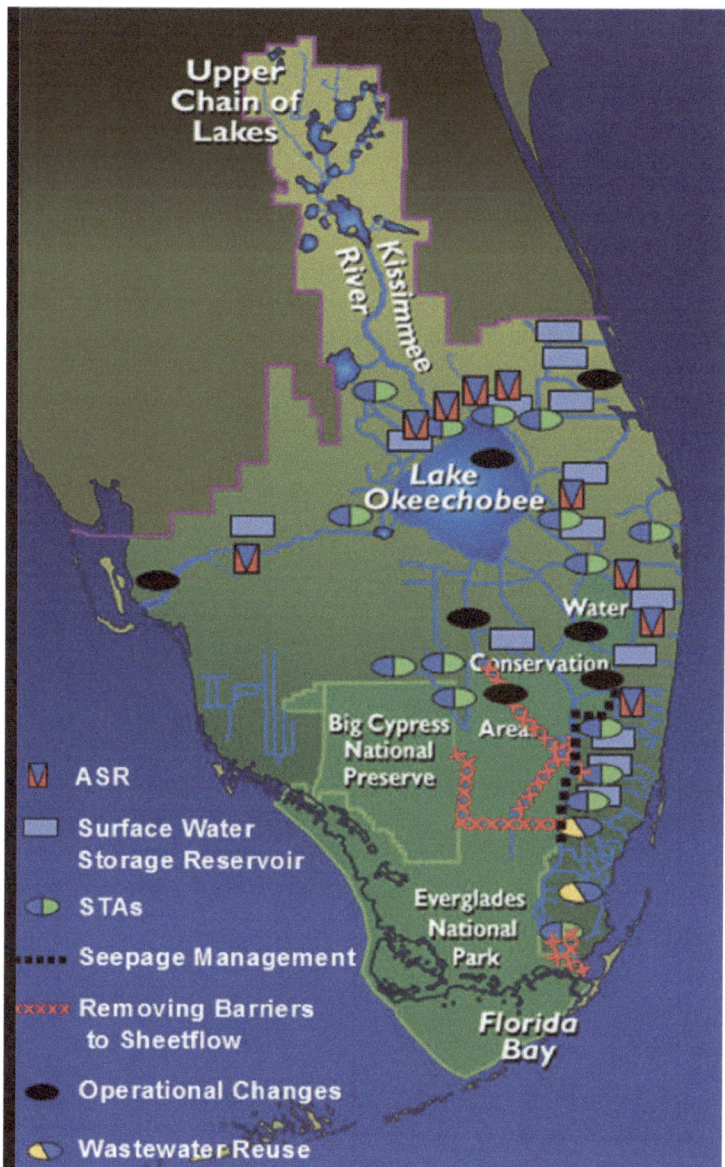

FIGURE 2-2 Elements of the Comprehensive Everglades Restoration Plan. The plan originally proposed a set of 68 projects, to be completed over an estimated 30 to 40 years, that were intended to achieve and sustain the essential hydrological and biological characteristics that defined the undisturbed ecosystem. A primary goal was to "get the water right" through a number of strategies including storing water underground for later use (Aquifer Storage and Recovery or ASR), creating surface water storage reservoirs, setting up natural stormwater treatment areas (STAs) to improve water quality, removing barriers to the historic sheet flows of water, altering operational plans to better attain CERP goals, and reusing wastewater. SOURCE: Courtesy of Laura Mahoney, U.S. Army Corps of Engineers.

the effects of restoration on these diverse system components. Many of the CERP projects address the connectivity that has been lost between parts of the system, and how actions taken in one part of the ecosystem influence conditions in other parts of the system. For example, a large project aims to increase the duration and magnitude of flows through the central Everglades into the northeast portion of Everglades National Park. However, increasing flows, even at levels that meet current water quality criteria, have been shown to adversely impact the biota, with losses of periphyton and increasing cattails evident in pilot testing. Thus, planners are grappling with ways to mitigate water quality impacts while maximizing flow benefits.

Finally, CERP projects are increasingly focused on resilience to expected future change, mostly as a result of climate change and sea-level rise. Impacts of sea-level rise can include flooding, shifts in extent and distribution of wetlands and other coastal habitats, and salinity intrusion into estuaries and groundwater systems. Increasing air temperatures reduce runoff and impact water availability, unless precipitation increases to counter these effects. CERP projects such as the South Florida Water Management District's new Sea Level Rise and Flood Resiliency Plan proposes infrastructure and nature-based solutions to address vulnerabilities to sea-level rise, storm surge, and extreme rainfall events.

Pandas, People, and Policies

This example encompasses all four themes presented above, taking a systems approach and using Wolong Nature Reserve in southwestern China as an illustration of terrestrial systems (Figure 2-3). The reserve was established in 1975 to conserve giant pandas, a global wildlife icon (Liu et al. 2016). It is within a global biodiversity hotspot (Myers et al. 2000). With an area of 2,000 km^2, it is home to several thousand plant and animal species, including approximately 150 (or 8% of the total population) wild giant pandas, and several thousand local residents (mainly farmers). Wolong constitutes a typical coupled human and natural system, where humans affect the natural systems (including panda individuals and habitat), which provide a variety of ecosystem services, such as water purification, carbon sequestration, and food (Yang et al. 2013). Wolong is vulnerable but is also resilient to disturbances such as the 2008 Wenchuan earthquake (Zhang et al. 2014). It has been a living laboratory of systems integration for understanding and managing coupled human and natural systems, including biological systems underpinning CSB, since 1996 (Liu et al. 2016).

Scaling has been a major research topic in Wolong and beyond. The spatial patterns of panda habitat based on data from satellites and field surveys vary at different scales. At a local scale (e.g., forest-stand patch), an entire area may be suitable or unsuitable. At a large scale, some areas are suitable habitats while other areas are unsuitable. The discovery in Wolong of faster growth in the number of households (a major driver behind biodiversity and habitat dynamics) than human population size led to the finding of a similar pattern at the global scale (Liu et al. 2003).

There are different or similar patterns across temporal scales. Using GPS (global positioning systems) collars on individual wild giant pandas, Zhang et al. (2015) inves-

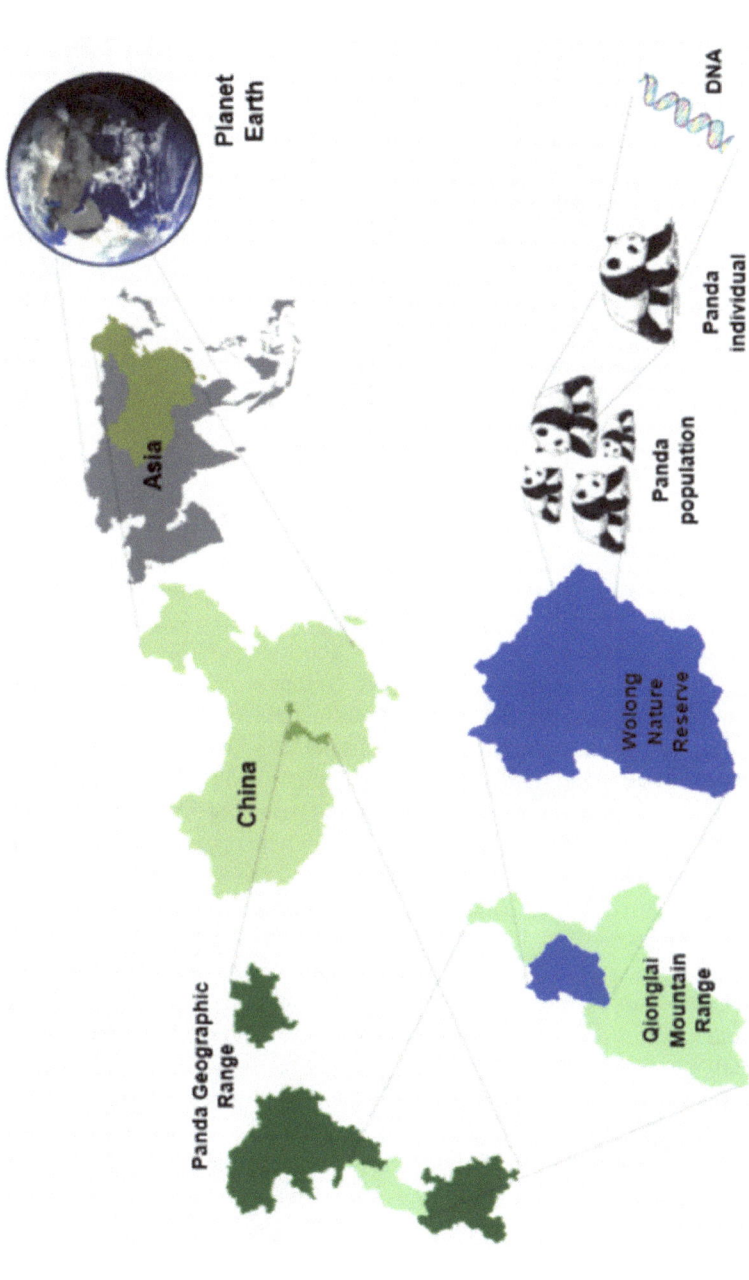

FIGURE 2-3 Research on pandas, people, and policies across scales from DNA to planet Earth. The maps were made using ArcGIS. The individual panda icon is from https://www.shutterstock.com/search/line-drawing-panda. The panda population icon was obtained by multiplying and transforming the panda individual icon. The DNA icon is from: https://commons.wikimedia.org/wiki/File:201812_DNA_double-strand_C.svg. The planet icon is from https://visibleearth.nasa.gov/images/57723/the-blue-marble.
SOURCE: Andrés Viña and Jianguo Liu.

tigated the daily and seasonal activity patterns. Contrary to the literature, most pandas were not crepuscular but showed three daily activity peaks, in the morning, afternoon, and around midnight. In terms of seasonal pattern, panda activity peaked in June, decreased in August and September, and increased again from November to March of the next year. For the total amount of panda habitat, the general temporal patterns of loss and recovery in 1976–2013 are similar in Wolong Nature Reserve (Liu et al. 2016) and across the entire geographic distribution range scale (Xu et al. 2017).

Emergent properties appear at higher organizational scales. Connor et al. (2023) used noninvasive fecal genetic (DNA) sampling to observe panda individuals in a wild population and infer association networks according to their spatiotemporal patterns. Even though the panda was thought of as a solitary species, the social network analyses revealed that cluster members preferred to associate with each other. This social clustering is an emergent property that does not show at the individual or DNA scales (Figure 2-3). These results suggest that many other "solitary" species may also have strong associations among individuals and emergent properties at higher organizational scales.

Multiscale interactions and feedback among pandas, people, and policies shape systems dynamics, according to the systems approach that integrates various sources of data (e.g., remote sensing, field investigations, tracking through global positioning system collars, socioeconomic surveys, government documents) and a portfolio of methods (e.g., DNA analysis, social network analysis, statistical analysis, systems modeling, agent-based modeling). Results indicate that humans affect panda habitat through activities such as farming and forest harvesting. Changes in panda habitat prompt the government to develop and implement new policies, and new policies change human activities, which in turn affect panda habitat across different scales (Liu et al. 2016).

Research on Wolong inspired the development of the metacoupling framework that integrates different types of human–nature interactions across space. From this perspective, Wolong is a focal system and interacts with adjacent systems such as neighboring counties where some women marry men in Wolong and move into Wolong, and with distant systems such as countries like the United States where many people visit Wolong as tourists. Conversely, Wolong may affect nearby areas where wild pandas move out of the reserve and distant areas where pandas in Wolong's breeding center are loaned to zoos such as the San Diego Zoo and the Smithsonian National Zoo in the United States.

Conservation efforts across local to global scales have led to the transition to sustainability. At the local scale, farmers in Wolong return cropland to forests, and reserve staff implement and monitor conservation efforts. Provincial and central governments, as well as international organizations on continents such as Asia, Europe, and North America, provide financial and technical support for the local farmers and reserve staff. Some members of the international and interdisciplinary team have also written blogs about the research for the general public. Science communicators and global news media outlets (e.g., BBC, *The New York Times*, *Time for Kids*, China's Xinhua News Agency) have widely reported the findings. These and other efforts have greatly enhanced stakeholder engagement as well as policies and practices. As a result, the collective action has transformed the habitat from long-term losses (even after the establishment of Wolong as a nature reserve) to recovery and the ultimate removal of the panda from the endangered species list of the International Union for Conservation of Nature in 2016. In this sense, panda habitat and population, as well as many ecosystem services, such as carbon sequestration, are on the trajectory to sustainability.

Methods and insights from Wolong have been applied to understanding and managing biodiversity as well as ecosystem services and other environmental challenges beyond Wolong, such as the Qionglai Mountain Range, Panda Geographic Range in China, other parts of Asia, and many other parts of the world (Figure 2-3). For example, the agent-based model developed for Wolong has been adapted to Chitwan National Park in Nepal (An et al. 2014). The habitat mapping methods developed in Wolong enabled studies on giant panda habitat dynamics across the species' geographic range (Viña et al. 2010). The work in Wolong also inspired efforts to detect changes in protected areas at the national, continental, and global scales (Yang et al. 2019, 2021). Comparisons between Wolong and other areas with different socioeconomic–ecological conditions around the world indicate that they share many complex attributes (e.g., time lags, nonlinearity, legacy effects, heterogeneity) (Liu et al. 2007).

The broad utilities of multiscale research on the four integrated themes demonstrated in this example suggest their potential value for studying other sites. These include sites of the Long-Term Ecological Research, International Long-Term Ecological Research, Long-Term Agroecosystem Research to NEON, as described in Chapter 4.

REFERENCES

Allen, T.F.H., and T.B. Starr. 1982. *Hierarchy: Perspectives for Ecological Complexity*. Chicago: University of Chicago Press.

An, L., A. Zvoleff, J. Liu, and W. Axinn. 2014. Agent-based modeling in coupled human and natural systems (CHANS): Lessons from a comparative analysis. *Annals of the Association of American Geographers* 104: 723-745. https://doi.org/10.1080/00045608.2014.910085.

Ashrafi, S., R. Kerachian, P. Pourmoghim, M. Behboudian, and K. Motlaghzadeh. 2022. Evaluating and improving the sustainability of ecosystem services in river basins under climate change. *Science of the Total Environment* 806 https://doi.org/10.1016/j.scitotenv.2021.150702.

Band, L.E., T. Hwang, T.C. Hales, J.M. Vose, and C.R. Ford. 2012. Ecosystem processes at the watershed scale: Mapping and modeling ecohydrological controls of landslides. *Geomorphology* 137:159-167. doi:10.1016/j.geomorph.2011.06.025.

Carlson, A.K., W.W. Taylor, and S.M. Hughes. 2020. The metacoupling framework informs stream salmonid management and governance. *Frontiers in Environmental Science* 8:27. doi:10.3389/fenvs.2020.00027.

Cavender-Bares, J.M., E. Nelson, J.E. Meireles, J.R. Lasky, D.A. Miteva, D.J. Nowak, W.D. Pearse, M.R. Helmus, A.E. Zanne, W.F. Fagan, C. Mihiar, N.Z. Muller, N.J.B. Kraft, and S. Polasky. 2022. The hidden value of trees: Quantifying the ecosystem services of tree lineages and their major threats across the contiguous US. *PLOS Sustainability and Transformation* 1(4):e0000010. https://doi.org/10.1371/journal.pstr.0000010.

Clark, W.C. and A. Harley. 2020. Sustainability science: Toward a synthesis. *Annual Review of Environment and Resources* 45:331-386. https://doi.org/10.1146/annurev-environ-012420-043621.

Clark, J.S., R. Andrus, M. Aubry-Kientz, Y. Bergeron, M. Bogdziewicz, D.C. Bragg, D. Brockway, et al. 2021. Continent-wide tree fecundity driven by indirect climate effects. *Nature Communications* 12:1242. https://doi.org/10.1038/s41467-020-20836-3.

Connor, T., K. Frank, M. Qiao, K. Scribner, J. Hou, J. Zhang, A. Wilson, V. Hull, R. Li, and J. Liu. 2023. Social network analysis uncovers hidden social complexity in giant pandas. *Ursus* (34e9):1-13. https://doi.org/10.2192/URSUS-D-22-00011.1.

Costanza, R., R. d'Arge, R. de Groot, S. Farber, M. Grasso, B. Hannon, K. Limburg, S. Naeem, R.V. O'Neill, J. Paruelo, R.G. Raskin, P. Sutton, and M. van den Belt. 1997. The value of the world's ecosystem services and natural capital. *Nature* 387:253-260. https://doi.org/10.1038/387253a0.

Cropper, M., J. Puri, and C. Griffiths 2001. Predicting the location of deforestation: The role of roads and protected areas in North Thailand. *Land Economics* 77(2):172-186. https://doi.org/10.2307/3147088.

Daily, G., and P.A. Matson. 2008. Ecosystem services: From theory to implementation. *Proceedings of the National Academy of Sciences of the United States of America* 105:9455-9456. https://doi.org/10.1073/pnas.0804960105.

Dakos, V., and S. Kéfi. 2022. Ecological resilience: What to measure and how. *Environmental Research Letters* 17(4): 043003. https://doi.org/10.1088/1748-9326/ac5767.

Davis, M.B., and R.G. Shaw. 2001. Range shifts and adaptive responses to Quaternary climate change. *Science* 292:673-679. https://doi.org/10.1126/science.292.5517.67.

DeFries, R.S., T. Rudel, M. Uriarte, and M. Hansen. 2010. Deforestation driven by urban population growth and agricultural trade in the twenty-first century. *Nature Geoscience* 3(3):178-181. https://doi.org/10.1038/ngeo756.

Dennis, B., and J. Koh. 2023. Smoke from Canadian wildfires engulfs East Coast, upending daily life. *The Washington Post*. https://www.washingtonpost.com/climate-environment/2023/06/07/air-quality-nyc-us-canada-wildfire-smoke/ (accessed March 12, 2024).

Diaz, S., J. Settele, E.S. Brondízio, H.T. Ngo, J. Agard, A. Arneth, P. Balvanera, et al. 2019. Pervasive human-driven decline of life on Earth points to the need for transformative change. *Science* 366(6471):eaax3100. https://doi.org/10.1126/science.aax3100.

Folke, C., Å. Jansson, J. Rockström, P. Olsson, S.R. Carpenter, F.S. Chapin III, A.-S. Crépin, et al. 2011. Reconnecting to the biosphere. *AMBIO* 40:719-738. https://doi.org/10.1007/s13280-011-0184-y.

Grimm, N.B., D. Foster, P. Groffman, J.M. Grove, C.S. Hopkinson, K.J. Nadelhoffer, D.E. Pataki, and D.P.C. Peters. 2008. The changing landscape: Ecosystem responses to urbanization and pollution across climatic and societal gradients. *Frontiers in Ecology and the Environment* 6(5):264-272. https://doi.org/10.1890/070147.

Hales, T.C., and C.F. Miniat. 2017. Soil moisture causes dynamic adjustments to root reinforcement that reduce slope stability. *Earth Surface Processes and Landforms* 42:803-813. https://doi.org/10.1002/esp.4039.

Hales, T.C., C.R. Ford, T. Hwang, J.M. Vose, and L.E. Band. 2009. Topographic and ecologic controls on root reinforcement. *Journal of Geophysical Research* 114: F03013. https://doi.org/10.1029/2008JF001168.

Healing, S., C.E. Benkwitt, R.E. Dunn, and N.A.J. Graham. 2024. Seabird-vectored pelagic nutrients integrated into temperate intertidal rocky shores. *Frontiers in Marine Science* 11. https://doi.org/10.3389/fmars.2024.1343966.

Heffernan, J.B., P.A. Soranno, M.J. Angilletta, Jr., L.B. Buckley, D.S. Gruner, T.H. Keitt, J.R. Kellner, et al. 2014. Macrosystems ecology: Understanding ecological patterns and processes at continental scales. *Frontiers in Ecology and the Environment* 12: 5-14. https://doi.org/10.1890/130017.

Holt, R.D. 1997. From metapopulation dynamics to community structure: Some consequences of spatial heterogeneity. Pp. 149-164 in *Metapopulation Biology: Ecology, Genetics, and Evolution*, I. Hanski and M.E. Gilpin, eds. San Diego: Academic Press.

Hou, W., T. Hu, L. Yang, X. Liu, X. Zheng, H. Pan, X. Zhang, S. Xiao, and S. Deng. 2023. Matching ecosystem services supply and demand in China's urban agglomerations for multiple-scale management. *Journal of Cleaner Production* 420:138351. https://doi.org/10.1016/j.jclepro.2023.138351.

Hwang, T., L. Band, T.C. Hales, C.F. Miniat, J.M. Vose, P.V. Bolstad, B. Miles, and K. Price. 2015. Simulating vegetation controls on hurricane-induced shallow landslides with a distributed ecohydrological model. *Journal of Geophysical Research: Biogeoscience* 120:361-378.

Ibáñez, I., L. Petri, D.T. Barnett, E.M. Beaury, D.M. Blumenthal, J.D. Corbin, J. Diez, J.S. Dukes, R. Early, I.S. Pearse, C.J.B. Sorte, M. Vilà, and B. Bradley. 2023. Combining local, landscape, and regional geographies to assess plant community vulnerability to invasion impact. *Ecological Applications* 33(4):e2821. https://doi.org/10.1002/eap.2821.

Inamine, H., A. Miller, S. Roxburgh, A. Buckling, and K. Shea. 2022. Pulse and press disturbances have different effects on transient community dynamics. *The American Naturalist* 200:571-583. https://doi.org/10.1086/720618.

Jaffe, D.A., S.M. O'Neill, N.K. Larkin, A.L. Holder, D.L. Peterson, J.E. Halofsky, and A.G. Rappold. 2020. Wildfire and prescribed burning impacts on air quality in the United States. *Journal of the Air & Waste Management Association* 70: 583-615. https://doi.org/10.1080/10962247.2020.1749731.

Leibold, M.A., M. Holyoak, and N. Mouquet, P. Amarasekare, J.M. Chase, M. Hoopes, R. Holt, J. Shurin, D. Tilman, M. Loreau, and A. Gonzalez. 2004. The metacommunity concept: A framework for multi-scale community ecology. *Ecology Letters* 7:601-613. https://doi.org/10.1111/J.1461-0248.2004.00608.X.

Levin, S.A. 1992. The problem of pattern and scale in ecology: The Robert H. MacArthur Award Lecture. *Ecology* 73:1943-1967. https://doi.org/10.2307/1941447.

Levin, S. 2024. Ecological resilience. *Encyclopedia Britannica*. https://www.britannica.com/science/ecological-resilience (accessed September 3, 2024).

Liu, J. 2023. Leveraging the metacoupling framework for sustainability science and global sustainable development. *National Science Review* 10(7):nwad090. https://doi.org/10.1093/nsr/nwad090.

Liu, J., G. Daily, P. Ehrlich, and G.W. Luck. 2003. Effects of household dynamics on resource consumption and biodiversity. *Nature* 421:530-533. https://doi.org/10.1038/nature01359.

Liu, J., T. Dietz, S.R. Carpenter, M. Alberti, C. Folke, E. Moran, A.N. Pell, et al. 2007. Complexity of coupled human and natural systems. *Science* 317:1513-1516. https://doi.org/10.1126/science.1144004

Liu, J., V. Hull, W. Yang, A. Viña, X. Chen, Z. Ouyang, and H. Zhang (eds.). 2016. *Pandas and People: Coupling Human and Natural Systems for Sustainability*. Oxford University Press. https://doi.org/10.1093/acprof:oso/9780198703549.001.0001.

Locey, K.J., and J.T. Lennon. 2016. Scaling laws predict global microbial diversity. *Proceedings of the National Academy of Sciences of the United States of America* 113(21):5970-5975. https://doi.org/10.1073/pnas.1521291113.

Loreau, M., N. Mouquet, and R.D. Holt. 2003. Meta-ecosystems: A theoretical framework for a spatial ecosystem ecology. *Ecology Letters* 6:673-679. https://doi.org/10.1046/j.1461-0248.2003.00483.x.

MacDonald, G., T. Wall, C.A.F. Enquist, S.R. LeRoy, J.B. Bradford, D.D. Breshears, T. Brown, et al. 2023. Drivers of California's changing wildfires: A state-of-the-knowledge synthesis. *International Journal of Wildland Fire* 32:1039-1058. https://doi.org/10.1071/WF22155.

Mann, M.E., Z. Zhang, M.K. Hughes, R.S. Bradley, S.K. Miller, S. Rutherford, and F. Ni 2008. Proxy-based reconstructions of hemispheric and global surface temperature variations over the past two millennia. *Proceedings of the National Academy of Sciences of the United States of America* 105(36):13252-13257. https://doi.org/10.1073/pnas.0805721105.

Martinez-Vilalta, J., and F. Lloret. 2016. Drought-induced vegetation shifts in terrestrial ecosystems: The key role of regeneration dynamics. *Global and Planetary Change* 144:94-108. https://doi.org/10.1016/j.gloplacha.2016.07.009.

Melillo, J.M., T.T. Richmond, and G. Yohe (eds.). 2014. *Climate Change Impacts in the United States: The Third National Climate Assessment*. U.S. Global Change Research Program. https://nca2014.globalchange.gov/report.

Miller, R.W. 2021. Thick smoke from western wildfires is traveling thousands of miles, clouding NYC skies. *USA Today*. https://www.usatoday.com/story/news/nation/2021/07/21/wildfire-smoke-cause-poor-air-quality-index-red-sun-nyc/8038354002/ (accessed March 8, 2024).

Myers, N., R. Mittermeier, C. Mittermeier, G.A.B. da Fonseca, and J. Kent. 2000. Biodiversity hotspots for conservation priorities. *Nature* 403:853-858. https://doi.org/10.1038/35002501.

NASEM (National Academies of Sciences, Engineering, and Medicine). 2023. *Progress Toward Restoring the Everglades: The Ninth Biennial Review—2022*. Washington, DC: The National Academies Press. https://doi.org/10.17226/26706.

Peters, D.P.C., P.M. Groffman, K.J. Nadelhoffer, N.B. Grimm, S.L. Collins, W.K. Michener, and M.A. Huston. 2008. Living in an increasingly connected world: A framework for continental-scale environmental science. *Frontiers in Ecology and the Environment* 6(5):229-237. https://doi.org/10.1890/070098.

Pickett, S.T.A., M.L. Cadenasso, and B. McGrath (eds). 2013. *Resilience in Ecology and Urban Design: Linking Theory and Practice for Sustainable Cities*. Dordrecht: Springer Science & Business Media.

Polasky, S., B. Bryant, P. Hawthorne, J. Johnson, B. Keeler, and D. Pennington. 2015. Inclusive wealth as a metric of sustainable development. *Annual Review of Environment and Resources* 40:445-466. https://doi.org/10.1146/annurev-environ-101813-013253.

Reid, R.S., M.E. Fernández-Giménez, and K.A. Galvin. 2014. Dynamics and resilience of rangelands and pastoral peoples around the globe. *Annual Review of Environment and Resources* 39:217-242. https://doi.org/10.1146/annurev-environ-020713-163329.

Remple, K.L., N.J. Silbiger, Z.A. Quinlan, M.D. Fox, L.W. Kelly, M.J. Donahue, and C.E. Nelson. 2021. Coral reef biofilm bacterial diversity and successional trajectories are structured by reef benthic organisms and shift under chronic nutrient enrichment. *NPJ Biofilms Microbiomes* 7:84. https://doi.org/10.1038/s41522-021-00252-1.

Rose, K.C., R.A. Graves, W.D. Hansen, B.J. Harvey, J. Qui, S.A. Wood, C. Ziter, M.G. Turner. 2016. Historical foundations and future directions in macrosystems ecology. *Ecology Letters* 20(2):147-157. https://doi.org/10.1111/ele.12717.

Seto, K.C., B. Güneralp, and L.R. Hutyra. 2012. Global forecasts of urban expansion to 2030 and direct impacts on biodiversity and carbon pools. *Proceedings of the National Academy of Sciences of the United States of America* 109:16083-16088. https://doi.org/10.1073/pnas.1211658109.

Shaffer, J.P., L.-F. Nothias, L.R. Thompson, et al. 2022. The Earth Microbiome Project 500 Consortium. Standardized multi-omics of Earth's microbiomes reveals microbial and metabolite diversity. *Nature Microbiology* 7:2128-2150. https://doi.org/10.1038/s41564-022-01266-x.

Shepherd, J.M., H. Pierce, and A.J. Negri. 2002. Rainfall modification by major urban areas: Observations from spaceborne rain radar on the TRMM satellite. *Journal of Applied Meteorology and Climatology* 41:689-670. https://doi.org/10.1175/1520-0450(2002)041%3C0689: RMBMUA%3E2.0.CO;2.

Stallard, R.F. 1998. Terrestrial sedimentation and the carbon cycle: Coupling weathering and erosion to carbon burial. *Global Biogeochemical Cycles* 12(2):231-257. https://doi.org/10.1029/98GB00741.

Tavárez, H., L. Elbakidze, O.J. Abelleira-Martínez, Z. Ramos-Bendaña, and N.A. Bosque-Pérez. 2022. Willingness to pay for gray and green interventions to augment water supply: A case study in rural Costa Rica. *Environmental Management* 69:636-651. https://doi.org/10.1007/s00267-021-01476-9.

Thompson, L.R., J.G. Sanders, D. McDonald, et al. 2017. The Earth Microbiome Project Consortium. A communal catalogue reveals Earth's multiscale microbial diversity. *Nature* 551:457-463. https://doi.org/10.1038/nature24621.

Turner, B.L., R.E. Kasperson, P.A. Matson, J.J. McCarthy, R.W. Corell, L. Christensen, N. Eckley, et al. 2003. A framework for vulnerability analysis in sustainability science. *Proceedings of the National Academy of Sciences of the United States of America* 100:8074-8079. https://doi.org/10.1073/pnas.1231335100.

Turner, B.L., E.F. Lambin, and A. Reenberg. 2007. The emergence of land change science for global environmental change and sustainability. *Proceedings of the National Academy of Sciences of the United States of America* 104:20666-20671. https://doi.org/10.1073/pnas.0704119104.

Turner, R.K., and G.C. Daily. 2008. The ecosystem services framework and natural capital conservation. *Environmental and Resource Economics* 39:25-35. https://doi.org/10.1007/s10640-007-9176-6.

UNEP (United Nations Environment Programme). 2018. *Inclusive Wealth Report: Measuring Sustainability and Well Being*. Nairobi, Kenya.

Viña, A., M.-N. Tuanmu, W. Xu, Y. Li, Z. Ouyang, R. DeFries, and J. Liu. 2010. Range-wide analysis of wildlife habitat: Implications for conservation. *Biological Conservation* 143:1960-1969. https://doi.org/10.1016/j.biocon.2010.04.046.

Vitousek, P.M. 2004. *Nutrient Cycling and Limitation. Hawai'i as a Model System*. Princeton, N.J.; Princeton University Press.

Wang, C., Y. Ye, and Z. Huang. 2024. Synergistic development in the Guangdong-Hong Kong-Macao Greater Bay Area: Index measurement and systematic evaluation based on industry-innovation-infrastructure-institution perspectives. *Journal of Cleaner Production* 434: 140093. https://doi.org/10.1016/j.jclepro.2023.140093.

Xiao, H., S. Bao, J. Ren, Z. Xu, S. Xue, and J. Liu. 2024. Global transboundary synergies and trade-offs among Sustainable Development Goals from an integrated sustainability perspective. *Nature Communications* 15:500. https://doi.org/10.1038/s41467-023-44679-w.

Xu, W., A. Viña, L. Kong, S.L. Pimm, J. Zhang, W. Yang, Y. Xiao, L. Zhang, X. Chen, J. Liu, and Z. Ouyang. 2017. Reassessing the conservation status of the giant panda using remote sensing. *Nature Ecology & Evolution* 1:1635-1638. https://doi.org/10.1038/s41559-017-0317-1.

Xu, Z., Y. Li, S.N. Chau, T. Dietz, C. Li, L. Wan, J. Zhang, L. Zhang, Y. Li, M.G. Chung, and J. Liu. 2020. Impacts of international trade on global sustainable development. *Nature Sustainability* 3:964-971. https://doi.org/10.1038/s41893-020-0572-z.

Yang, W., T. Dietz, W. Liu, J. Luo, and J. Liu. 2013. Going beyond the Millennium Ecosystem Assessment: An index system of human dependence on ecosystem services. *PLoS ONE* 8(5):e64581. https://doi.org/10.1371/journal.pone.0064581.

Yang, H., A. Viña, J.A. Winkler, M.G. Chung, Y. Dou, F. Wang, J. Zhang, Y. Tang, T. Connor, Z. Zhao, and J. Liu. 2019. Effectiveness of China's protected areas in reducing deforestation. *Environmental Science and Pollution Research* 26:18651-18661. https://doi.org/10.1007/s11356-019-05232-9.

Yang, H., A. Viña, J.A. Winkler, M. G. Chung, Q. Huang, Y. Dou, W.J. McShea, M. Songer, J. Zhang, and J. Liu. 2021. A global assessment of the impact of individual protected areas on preventing forest loss. *Science of the Total Environment* 777:145995. https://doi.org/10.1016/j.scitotenv.2021.145995.

Zhang, J., V. Hull, J. Huang, W. Yang, S. Zhou, W. Xu, Y. Huang, Z. Ouyang, H. Zhang, and J. Liu. 2014. Natural recovery and restoration in giant panda habitat after the Wenchuan earthquake. *Forest Ecology and Management* 319:1-9. https://doi.org/10.1016/j.foreco.2014.01.029.

Zhang, J., V. Hull, J. Huang, S. Zhou, W. Xu, H. Yang, W.J. McConnell, R. Li, D. Liu, Y. Huang, Z. Ouyang, H. Zhang, and J. Liu. 2015 Activity patterns of the giant panda (*Ailuropoda melanoleuca*). *Journal of Mammalogy* 96(6):1116-1127. http://dx.doi.org/10.1093/jmammal/gyv118.

Zhao, Z., M. Cai, F. Wang, J.A. Winkler, T. Connor, M.G. Chung, J. Zhang, H. Yang, Z. Xu, Y. Tang, Z. Ouyang, H. Zhang, and J. Liu. 2021. Synergies and tradeoffs among Sustainable Development Goals across boundaries in a metacoupled world. *Science of the Total Environment* 751:141749. https://doi.org/10.1016/j.scitotenv.2020.141749.

3

Theoretical Underpinnings for a Continental-Scale Biology

"Biological research is in crisis ... technology gives us the tools to analyse organisms at all scales, but we are drowning in a sea of data and thirsting for some theoretical framework with which to understand it ...." "[W]e now have unprecedented ability to collect data about nature.... You might say that we could in principle make an atom-by-atom description of what there is in nature, but there is now a crisis developing in biology.... ""that completely unstructured information does not enhance understanding. What people want is to understand, which means you must have a theoretical framework in which to embed this.... [P]eople who just collect data are not doing science in that sense."

Sydney Brenner – Interview. NobelPrize.org. Nobel Prize Outreach. Interview with the 2002 Nobel Laureates in Physiology or Medicine, Sydney Brenner, John E. Sulston, and H. Robert Horvitz, by science writer Peter Sylwan, December 12, 2002. https://www.nobelprize.org/prizes/medicine/2002/brenner/interview/.

OVERVIEW AND PROBLEM STATEMENT

The urgency to advance the science of the biosphere has never been more critical (Folke et al. 2021). With limited resources, capacity, and time for intervention, action, and conservation, there is a pressing need for more precise predictions to enhance efficiency. While prediction is integral to understanding, it is the more immediate goal for society and application (Potochnik 2020). Furthermore, the biosphere faces considerable uncertainty regarding the potential shifts resulting from policies that might advocate radical new interventions beyond traditional carbon emission reductions and conservation techniques. This scenario demands a deep, fundamental understanding to assess the effectiveness and potential consequences of any innovative course of action.

Theory is essential for advancing and shaping continental-scale biology (CSB). Gaps in theory in biology (NRC 2008) limit our ability to comprehensively define the scope and boundaries of CSB and to refine the underlying principles of CSB science. Moreover, a robust theoretical framework is crucial to effectively guide the search for and discern valuable insights from the vast influx of data and to moderate the increasing dependence on artificial intelligence (AI) and statistical forecasting, which we worry can lack principled approaches (Coveney et al. 2016; Enquist et al. 2024.) There is a pressing need for new initiatives to forge an integrative theory that spans the disciplines within CSB. Such a theory would not only synthesize and harmonize existing theoretical frameworks but also enhance our understanding and management of complex biological systems from genes to the biosphere.

A Grand Challenge—Forecasting the Future of the Biosphere

Box 1-2 describes as one of the characteristics of CSB that it "inherently incorporates multiple scales, from the subcellular to the global biosphere (Figure 1-1), from the local to global spatial extents, from less than a second to millennia." As emphasized by Harrison et al. (2021), our current ability to forecast the future of the biosphere and model its dynamics is, to put it mildly, challenged. The significant shortfalls in modeling the functioning of the biosphere as a core component of the climate system highlights significant gaps in our biosphere theory. Despite successes, substantial hurdles persist, particularly in reproducing large-scale phenomena. Both Earth system models (ESMs) and dynamic global vegetation models struggle with accuracy, failing to capture the amplification of the high-latitude seasonal cycle of atmospheric CO_2 (Grave et al. 2013, Thomas et al. 2016) and the relationship between the $^{13}C/^{12}C$ stable isotope ratio of atmospheric CO_2 and global land–atmosphere carbon exchange (Peters et al. 2018).

Persistent discrepancies in models over the effects of global warming on primary production, vegetation responses to precipitation changes, and the influence of CO_2 and nutrient availability (Ciais et al. 2013, Huntzinger et al. 2017, Wieder et al. 2015) have been highlighted for decades (Friedlingstein et al. 2006, VEMAP 1995) and were notably mentioned in the IPCC Fifth Assessment Report (IPCC, 2013). Recent studies confirm these ongoing issues (Arora et al. 2020).

Significant fundamental uncertainties still plague our understanding of the biosphere's responses to environmental changes. There remains considerable disagreement and uncertainty about how the biosphere, including its interactions with human activities, will react to increased atmospheric CO_2 levels and subsequent rises in ambient temperatures (Arora et al. 2020, Friedlingstein et al. 2006, 2014). This uncertainty extends to several critical areas: the existence and thresholds of specific ecological tipping points (Chaparro-Pedraza and de Roos 2020, Ditlevsen and Johnsen 2010, Drake et al. 2020, Dudney and Suding 2020, Lenton 2013), the actual trends in global biodiversity and whether it is truly decreasing (Dornelas et al. 2014, McGill et al. 2015), the rate and implications of species extinctions in the Anthropocene (Ceballos and Ehrlich 2018, Rothman 2017), and the long-term effects of geoengineering initiatives such as iron fertilization (Keith 2021). These failures and uncertainties indicate a pressing need to reassess and potentially overhaul the assumptions and methodologies used in current vegetation models. Developing new or significantly improved theoretical frameworks to enhance their predictive capabilities is crucial, necessitating a multidisciplinary approach that incorporates broader ecological and evolutionary insights. Such advancements are essential to effectively tackle the interlinked challenges of climate change, biodiversity conservation, and ecosystem management, propelling CSB forward.

Background

Science seeks to enhance our comprehension of the natural world, thriving on the dynamic interplay and tension between induction and deduction, and the balance between empiricism and theoretical frameworks. Observational and experimental data offer insights into the structure and functionalities of the natural phenomena around us. Theory, at its core, reflects our attempt to understand biological and physical phenomena and is used to classify, interpret, and predict. As this chapter's epigraph notes, a theoretical framework is more than a collection of facts or data. It is a logical framework developed for understanding and interpreting observations and facts. A previous report (NRC 2008) defines theory as "integral to each specific kind of scientific activity, including experimentation, observation, exploration, description, and technology development as well as hypothesis testing" and differentiates "facts and data," as distinct from theory, which are the structures that explain and interpret data and facts. Theory may evolve with new data, but facts/data do not."

Theory unifies disparate observations and empirical laws under a single conceptual umbrella, serving as a compass for scientific exploration. Indeed, as Marquet et al. (2014) note, "[t]heory reduces the apparent complexity of the natural world, because it captures essential features of a system, provides abstracted characterizations, and makes predictions for as-yet unobserved phenomena that additional data can be used to test. . . Data gathered through observation and experimentation provide clues about the structure and function of the natural world, and theory organizes existing data and new ideas into a cohesive conceptual framework to both explain existing observations and make novel predictions."

Theory enables us to classify, interpret, and predict observations and natural phenomena. It integrates various aspects of a phenomenon, offering a coherent narrative that elucidates underlying principles. By explaining observations as parts of a greater whole, theory guides our expectations under specific conditions. It is inherently dynamic, evolving with new evidence or contradictory findings. Theories are inherently predictive, setting the stage for empirical testing and verification.

The initial focus is to develop simple, tractable, mechanistic theories with relatively few variables and parameters. These may be caricatures of the system, but they play a crucial role because they attempt to incorporate the important variables and essential features that determine the system's organization, structure, development, and dynamics (Servedio et al. 2014). In Isaac Newton's Untitled Treatise on Revelation (section 1.1) he states, *"choose those constructions which without straining reduce things to the greatest simplicity …. Truth is ever to be found in the simplicity, and not in the multiplicity and confusion of things."* In his *Mathematical Principles of Natural Philosophy* (1687), his Rule I is "*No more causes of natural things should be admitted than are both true and sufficient to explain their phenomena.*" Ecology has a long history of building theories (Scheiner and Willig, 2011); the power of a parsimonious approach to theory development is that it is typically falsifiable so that models can be appropriately modified when their predictions are confronted with data. As Enquist et al. (2024) note, "A well-defined theory with specific testable predictions proven wrong by confrontation with data can provide important insights for moving a field in the right direction."

The power of a parsimonious approach to theory development is that it is typically falsifiable. The emphasis is on making these models falsifiable so that they can be appropriately modified when their predictions are confronted with data. Ecology has a long history of building theories (Scheiner and Willig, 2011); a well-defined theory with specific testable predictions proven wrong by confrontation with data can provide important insights for moving a field in the right direction.

There is a need to develop a core body of theory in CSB based on zeroth-order frameworks and first principles rooted in the foundational laws of biology, physics, and chemistry. A "zeroth-order framework" is the use of simplifying assumptions to first give a rough approximation to the solution to a problem. The most basic, essential factors are first considered and more complex or minor influences are ignored. This is typically the starting point in solving complex problems, allowing a scientist to capture the primary behavior of a system with minimal computational or analytical complexity (West 2017). Successive iterations, or orders of approximation, include increasingly more influences on the solution (e.g., more starting assumptions) that hopefully will successively better refine the approximation of the truth. Using the principle of parsimony, starting assumptions and input need to be chosen carefully. To quote the biologist J.B.S. Haldane, "*In scientific thought we adopt the simplest theories which will explain all the facts under consideration and enable us to predict new facts of the same kind*" (Haldane 1927).

A zeroth-order framework contains the building blocks of foundational assumptions or axioms from which a theory is built. These principles are typically so fundamental that they are generally accepted without needing empirical evidence. Building theory starting with a zeroth-order framework ensures a robust, logically coherent framework that can be universally applied and tested, and theory can be systematically built and expanded to include more complex and specific phenomena. Next, first principles are derived from zeroth-order assumptions but are often more specific and can be proved or derived through reasoning and logical deduction. "*First principles are the bedrock of science—that is, quantitative law-like postulates about processes underlying a given class of phenomena in the natural world with well-established validity, both theoretical and empirical (i.e., core knowledge)*" (Marquet et al. 2014, p. 703). In physics, using first principles often involves starting with basic laws like Newton's laws of motion or the laws of thermodynamics and applying them to work out complex phenomena from these fundamental truths. In biology, first principles are fundamental concepts or foundational truths from which more complicated theories and models are derived (e.g., the principles of homeostasis, stoichiometry, evolution by natural selection, conservation of energy, trade-offs in resource allocation, principles of exponential growth and carrying capacities, and trophic structure). Each of these principles provides a foundational framework from which more detailed and specific scientific inquiries and hypotheses can be constructed, explaining complex biological phenomena across different scales and contexts. "The ultimate goal is to develop quantitative, predictive theories grounded in underlying principles and supported by data, observation, and experimentation" (Enquist et al. 2024).

There is increasing urgency to address many significant biosphere challenges (Box 3-1) that directly threaten human well-being and socioeconomic stability (Díaz et al. 2019, McMichael 2013, Ruckelshaus et al. 2020, WMO 2021). The ability to accurately forecast biodiversity and ecosystem functioning, predict the onset and extent of future pandemics, or determine when the Amazon rainforest might hit a catastrophic tipping point demands our attention. Identifying the critical parameters and dynamics that drive these threats is essential for developing quantitative strategies to minimize adverse outcomes and mitigate potential disasters. However, the inherent complexity of the biosphere, characterized by biological processes that span a broad spectrum of spatial and temporal scales, presents a formidable challenge.

Such a coherent theoretical framework would help guide observations and experiments and will enable scientists to understand mechanisms of climate change, land use, and other significant aspects of CSB. Nature-based strategies, recent efforts to protect and improve the natural and enhanced environment by addressing biodiversity challenges and assessing climate mitigation strategies, could be improved by developing additional tools, including theories (Novick et al. 2024).

Lessons can be learned from the effectiveness of a theory-driven approach in climate and atmospheric sciences (Emanuel 2020, Enquist et al. 2024), which has proven highly effective in predicting climate patterns, understanding atmospheric processes, and forecasting future climate scenarios. Such a theory-based framework is an example that CSB could strive to emulate to predict changes in the biosphere amidst complex environmental challenges. The Earth sciences provide a compelling example, having anticipated the unfolding climate crisis by blending fundamental theories with progressively more sophisticated observations and experiments within a theory-driven simulation framework. Atmospheric and ocean science has not only accurately predicted global temperature changes but has also provided increasingly detailed projections of past, present, and future shifts in temperature and circulation patterns (Arias et al. 2021, Hausfather et al. 2020). In contrast, similar efforts to predict corresponding phenomena within the biosphere have not progressed as rapidly (e.g., compare and contrast discussions in Doak et al. 2008, Fisher et al. 2018, Moorcroft 2006). This comparison underscores the need for CSB to challenge past approaches and research investments to better adopt and adapt these successful theory-based approaches to better anticipate and respond to biospheric changes.

CHALLENGES IN DEVELOPING CSB THEORIES

The committee identified three challenges in developing CSB theories. First, a major challenge is the explosion in our capacity to amass extensive datasets of different parameters, detail life's rich biology at multiple scales, map the diversity of life, and

> **BOX 3-1**
> **Role of Theory in the Core Themes of CSB**
>
> As described in Chapter 2, four core themes underpin CSB: biodiversity and ecosystem function, resilience and vulnerability, connectivity, and sustainability of ecosystem services. Theory plays a role in each of these themes. For example, *biodiversity and ecosystem function* includes a role for developing theories to understand the relationships between biodiversity and function across scales and how emergent properties at one scale influence those at the next. *Resilience and vulnerability* needs theories to help researchers understand how biodiversity influences resilience, how to identify potential tipping points in systems, and how resilience and vulnerability at different scales influence each other, at either finer or coarser scales. Theory is also required to support the advancement of metacoupling analysis, described in Chapter 2 as a key tool for understanding *connectivity* among natural and human systems across multiple scales. Similarly, maintaining the *sustainability of ecosystem services* requires theories to guide research on the linkages between natural ecosystems and the services they provide to people.

reach a granularity that was once unthinkable. However, as Brenner underscores in this chapter's epigraph, the sheer quantity of data, even if atom-by-atom, does not automatically translate to an enhanced understanding of biological phenomena. Currently, the unprecedented levels of detail that can be seen at any scale are far outpacing the derivation of theoretical frameworks that can process this information across multiple scales and glean understanding and predictive ability.

Second, multiscale biological research is challenged by the need to go from detail to insight. Much advancement has been made in our ability to map and describe the intricate details of biological and ecological processes across scales (Hampton et al. 2013). A theoretical framework is vital because it provides the necessary abstraction that allows for meaningful interpretation, prediction, and utility in scientific endeavors and applications. In essence, science is the delicate balance between abstraction and synthesis (theory) and the retention of minimal, but sufficient detail for a theory to be effective in science.

Third, the lag in developing predictive CSB science is influenced not only by the complexity of biology or data scarcity but also by the tensions between the way scientists gather and interpret data as characterized by three predominant scientific cultures within biology (Enquist et al. 2024). The "variance culture" or natural history, focuses on detailed observation and cataloging, such as bird counts and collecting flora and fauna, emphasizing the meticulous documentation of biological diversity. It is arguably the basis of modern biology, including molecular biology, genetics, and numerous fields that currently do not rely much on natural history. The emphasis is on detailed observations and focuses on differences and deviations of taxa, clades, and specific locations or regions. It leans more toward experimental and observational methods of investigation. The "exactitude culture" advocates for highly detailed models to mimic real-world complexity, often prioritizing precision over practical scalability and emphasizes incorporating more detail, typically focusing on specific problems or phenomena and more general concepts or understanding. Approaches include detailed statistical modeling, machine learning (ML), and AI untethered to parsimony and assessing competing models based on information criteria. Models tend to be phenomenological with many parameters, often disconnected from underlying principles. Conversely, the "coarse-grained culture" prioritizes simplification and general principles, aiming to distill complex information into overarching insights. "This approach can include mathematical derivations of probabilistic outcomes or take the form of a parsimonious statistical theory" (Enquist et al. 2024). The perspectives from each scientific culture, while valuable, often operate in isolation within the biosciences. Progress in tackling complex problems and advancing CSB emerges when these diverse approaches are integrated, combining detailed empirical data with high-level theoretical synthesis to foster a comprehensive understanding essential for addressing global ecological challenges.

These three challenges are further intensified by the climate and biodiversity crises. The escalating climate crisis, the alarming rates of species extinction, and the urgent need to preserve biodiversity underscore the necessity for a predictive science of the biosphere and to help guide issues central to the UN Climate Change Conference COP25 agenda and the Intergovernmental Science-Policy Platform on Biodiversity and Ecosystem Services. The ability to not just collect but also to guide the collection and distillation of vast biological data, meaningfully interpret, obtain knowledge, and use this information to then make predictions is needed.

Theories are needed that not only bridge the micro and macro scales of life but also provide actionable insights into mitigating climate change impacts, conserving biodiversity, and enhancing ecosystem services. In this context, the balance between detailed data collection and theoretical abstraction is not just a scientific endeavor; it's a requisite tool for addressing some of the most pressing ecological and societal challenges of our time. As discussed below, while several elements of theoretical strands are in place for CSB there are several notable gaps.

REQUIREMENTS OF CSB THEORIES

General theories for CSB requires bridging gaps in our current understanding of both small- and large-scale biological and ecological processes. For example, climate change affects ecosystems through altered mean conditions and increased variability (Turner et al. 2020) alongside rising atmospheric carbon dioxide levels. These changes exacerbate other ecological pressures such as habitat loss and degradation, defaunation, and fragmentation. As emphasized by Malhi et al. (2020), understanding the ecological dynamics of these impacts, pinpointing hotspots of vulnerability and resilience and identifying effective management interventions are essential for enhancing biosphere resilience (Jung et al. 2021, Molotoks et al. 2020). Moreover, ecosystems and biodiversity play a crucial role in both mitigating and adapting to climate change (Morecroft et al. 2019). Increasingly CSB is being called to assess how ecosystem management and restoration have the potential to contribute nature-based solutions to address both the causes and consequences of climate change. However, the effectiveness, scalability, and magnitude of different nature-based strategies need to be explored, better understood, and evaluated (Bennett et al. 2016, Malhi et al. 2020). Exploring and quantifying the mechanisms, potential, and limits of nature-based solutions are vital for informed decision making and policy formulation (Malhi et al. 2020, Morecroft et al. 2019).

Processes that define CSB encompass multiple dimensions of biodiversity, including genomic, taxonomic, functional, and phylogenetic measures (Cadotte et al. 2009, Naeem et al. 2012, 2016). CSB theory needs to provide a conceptual framework upon which multiple dimensions of biodiversity can be linked to multiscale (including molecular, cellular, organismal, and ecological) data, recognizing that the framework may change as new technologies and knowledge evolve. For example, theories can be applied to a variety of questions spanning everything from research on invasive species dynamics, biodiversity responses to various drivers, mass extinction, pandemics, to the spatial variation of ecosystem stocks and flows. However, an integrative cycle between observation, data analysis, and theory development (see Box 3-2) would enable linkages between biosphere predictions and would foster the development of new approaches and technologies to address critical ecosystem and societal needs.

CSB theories, present and future, will necessarily range from targeted (e.g., trait-based theories of carbon sequestration) to global in utility and function (e.g. projecting geographic variation of extinction risks to differing climate and human drivers). Below, the committee provides a catalog of current CSB-relevant theories that embody one or more of the following core attributes for developing effective CSB theories. These core attributes are:

a. **Scaling and Multiscale Application:** CSB theories need to be applicable at various spatial and temporal scales. These scales encompass attributes of individuals (traits, genes), populations, and species assemblages on landscapes, to ecosystem functions and up to the entire biosphere. CSB theories can provide solutions that navigate cross-scale questions and consider all the dimensions of biodiversity, including genomic, taxonomic, functional, and phylogenetic diversity, and include ecosystem pools and fluxes.
b. **Data Guidance and Technological Integration:** Modern technology and informatics offer unprecedented monitoring capabilities for biosphere processes. However, integration of data across scales for theory application and development remains a challenge. Theory is needed to inform the collection and management of big data by playing a pivotal role in ensuring that data collection aligns with the needs of understanding multiscale biological and ecological processes. Theoretical frameworks for CSB need to mesh and help guide the collection of diverse datasets, technologies, and monitoring programs and will help in interpreting which of their outputs is needed.
c. **Identify and Amplify Biological Linkages:** CSB theories also need to forge connections across biological processes and scales from the molecular and cellular levels to populations, entire ecosystems, and the biosphere (Figure 1-1).

d. **Comprehensive Scope:** CSB theories need to unify disparate functional processes from microbial to ecological and physiological processes, including material flux through the biosphere.
   e. **Human Aspect Theories Affecting Study Sites, Experimental Design That Might Bottleneck Theory Development.** Humans make decisions on experimental design, site selection, hypotheses to be examined, and how data are analyzed, which can potentially add bias, which needs to be evaluated prior to study initiation. Encompassing different perspectives, experiences, and knowledge foundations of research teams can drive innovation and analyses for all aspects of CSB.
   f. **Inclusion of a Metacoupling Framework in Theory Development That Synthesizes Human Social and Ecological Interactions Across Scales.** Humans also affect the function of the biosphere, but too often their effects on these processes are not fully captured in experimental designs and ultimately, in theory. As described in Chapter 2, metacoupling analysis is an emerging effort to define and address connectivity between human influences and natural ecosystems near and far. However, capturing these processes and biases in theory development to inform experimental questions is a challenge.

The cyclical process of theory development is intrinsic to the progress of CSB (Box 3-2). It emphasizes that theories are never static but continually evolving entities based on hypothesis testing and new data. Assumptions and predictions are not rigid constructs but flexible tools that adapt and grow as our understanding deepens. They facilitate a dynamic dialogue between theories and empirical evidence, driving the field forward and providing a structured approach to complex biological inquiry. Theory helps determine what data are crucial, what to measure, and where to focus attention. It synthesizes data by highlighting connections and predicts unmeasured aspects.

Select CSB-Relevant Theories

In reviewing the landscape of biological theories that is relevant for CSB, the committee focused on a sample of theories that provide a framework for integration across spatial, organizational, and temporal scales. The committee also considered the theory's ability to assist in interpreting empirical data, moving us closer to a more integrative, quantitative, and predictive understanding of biological systems.

Neutral Theory of Biodiversity

The neutral theory of biodiversity (NTB) provides a framework for understanding the spatial and temporal dynamics of biodiversity via the impact of stochastic demographic processes on community structure and dynamics across ecological to macroevolutionary timescales (Hubbell 2001, Rosindell et al. 2012). NTB generates a broad array of predictions concerning phenomena such as species abundance distributions, species–area relationships, phylogenetic tree structures, and correlations between species

> **BOX 3-2**
> **Developing, Testing, and Advancing Theory**
>
> The scientific triad of theory, observation, and experiment forms the critical pillars of the scientific method as illustrated by the ecological forecasting cycle (Figure 3-1-1). This triad serves as the guiding force in CSB. Its significance to CSB science is central.
>
> - *Start with and Build on Assumptions:* The development of any theory begins with assumptions. These are fundamental premises or generalizations that underpin the theory. They serve as the starting point and framework for understanding complex biological phenomena. The process begins by identifying zeroth-order principles and building on assumptions. The role of parsimony, or simplicity, is vital in theory development, aiming for the most straightforward and minimal explanations.
> - *Generating Predictions:* Based on the underlying assumptions, theories generate specific predictions. These predictions provide concrete, testable statements about what we expect to observe in the natural world. This stage provides critical insights into the validity of the theories and uncovers areas where the assumptions may be oversimplified or incorrect.
> - *Testing—Accepting, Rejecting, or Improving Assumptions and Predictions:* The cyclical nature of theory development demands constant evaluation. Assumptions and predictions may be accepted, rejected, or improved upon. Improvement often involves incorporating new data, adjusting to emerging insights, and aligning the theories with current scientific understanding.

richness and macroevolutionary rates of speciation and extinction. With remarkably few variables, NTB posits that niche differences among species do not significantly influence their ecological success, effectively treating species as demographically equivalent in terms of individual rates of speciation, birth, death, and dispersal (Volkov et al. 2005).

NTB's applicability extends across various scales of biological organization, from microbial communities to global biodiversity patterns, making it particularly relevant to CSB. It offers predictions on the distribution of species commonness, rarity, and the temporal dynamics of biodiversity change (Hubbell 2001), serving as a foundational null model to establish baseline expectations. Although many biodiversity patterns diverge from these neutral expectations, deviations are analytically valuable. For instance, deviations in studies like those analyzing human microbiome datasets reveal that the host environment significantly shapes community composition and assembly (Li and Ma 2016), thus providing critical insights into the specific ecological and evolutionary mechanisms influencing biodiversity.

- **Reiteration of the Cycle:** The process then repeats, with the refined assumptions and predictions undergoing further development, testing, and improvement. The cycle is an ongoing, iterative process that ensures theories evolve with scientific advancements and remain relevant and robust.

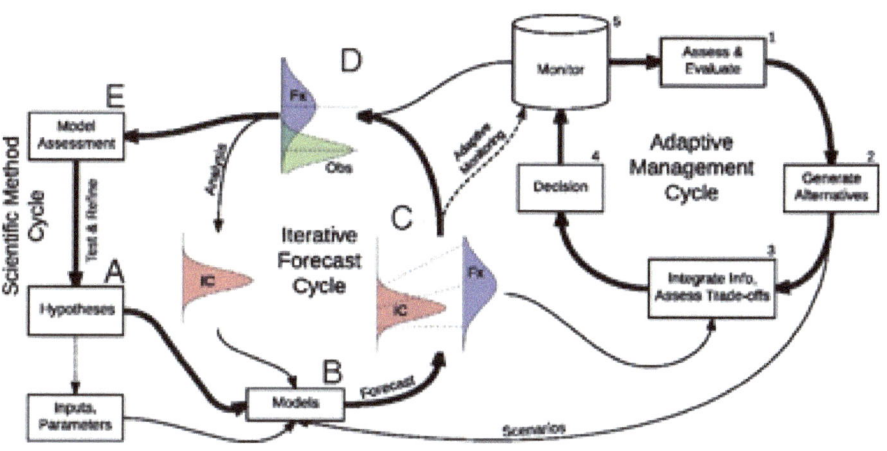

FIGURE 3-1-1 Near-term ecological forecasting cycle.
SOURCE: Dietze et al. 2018.

Phenotypic Optimality from Ecoevolutionary Optimality

Physiological optimality models offer an alternative to leverage current ecoevolutionary optimality (EEO) theory to understand how plants optimize traits in response to environmental pressures. EEO theory posits that natural selection eliminates less competitive traits, allowing plants to adjust their physiological responses across timescales from days to millennia for optimal survival and reproduction. The theory hinges on the mechanistic links between plant functional traits, resource acquisition, and biogeochemical cycling, impacting plant competitiveness (Franklin et al. 2020, Harrison et al. 2021).

EEO models offer parsimonious, testable parameters that encapsulate critical trade-offs, such as maximizing carbon gain while minimizing water loss. This approach has successfully been applied to predict vegetation patterns across climates, evidenced in both natural and agricultural ecosystems (Qiao et al. 2020, 2021; Yang et al. 2018). Enhanced by the wealth of data from plant trait databases and satellite remote sensing, these models can rigorously test the universal patterns and simulate ecosystem responses to environmental drivers (Wieder et al. 2015).

Advancements in EEO highlight its potential to improve vegetation models by focusing on mean phenotypes adapted to specific climates, thereby predicting observable shifts in traits at leaf, plant, and community levels (Smith et al. 2019, Wang et al. 2020). Foundational physiological models such as the Farquhar, von Caemmerer, and Berry (FvCB) model underpin these insights, quantifying the physiological trade-offs in photosynthesis (Collatz et al. 1991, Farquhar et al. 1980). Integrating these EEO modules into broader modeling and scaling frameworks could significantly advance our understanding of plant community adaptations to environmental changes over multiple timescales, aligning with the goals of CSB.

Niche Theory

Niche theory offers a conceptual framework for understanding how species persist and interact within their environments. Niche theory is relevant to CSB because it explores the interplay between biogeography and environmental conditions on the persistence and abundance of species and that can drive population genomics (Vandermeer 1972). Central to this framework are "consumer-resource models," which partition ecosystems into two main categories: resources, such as sunlight or soil moisture, and consumers, including all living organisms. Niche theory attempts to define the range of possible relationships between the two groups, for example, competition for food and predator–prey relationships (Abrams 1986). Niche theory has evolved to include computational models that consider various ecological rules and trade-offs that can predict species abundances and distributions based on stabilizing and equalizing mechanisms (Leibold 1995).

Niche theory provides a basis to forecast the distribution of species and biodiversity on the planet (Guisan and Thuiller 2005, Thuiller et al. 2008). It also provides a foundation on biogeographical and environmental drivers for organismal reproduction and survival and has been used to make predictions for species abundances under different environmental parameters. Examples include how temperature, precipitation, and soil pH shape microbial communities and their interactions with plants (Kivlin et al. 2021). Niche concepts and theory, in the form of "ecological niche models" or "bioclimatic envelope models" have become central in efforts to understand how future climate change may have an impact on species and their habitats (Guisan and Zimmermann 2000, Letten et al. 2017, Morin and Lechowicz 2008). While arguably overly statistical in practice, species distribution models can be based on physiological and biological mechanisms (unimodal responses to climate gradients, including dispersal limitation and thermal and other physiological limits on distribution). Current ecological niche constraints are used to project future species distributions under environmental change (Pillet et al. 2022). Ecological niche models use information on environmental features that define the current ecological niche of a species in association with the future distributions of those features derived from climate-change models to project where the species' niche requirements may be satisfied in the future.

Metabolic Scaling Theory

Scaling relationships are observed at multiple levels of biology. Building on the allometric and metabolic rules of life, metabolic scaling theory (MST) offers a theoretical framework for understanding the origin of these scaling relationships. MST also offers a unified approach to scaling up from cells to ecosystems to large-scale biological phenomena. MST is relevant to CSB because it provides a framework for investigating and assessing the interplay between biological processes, including metabolism and size of an organism, which can be scaled up to the level of populations and ecosystems (West et al. 1997, 1999). MST is a set of related theoretical applications of the scaling of metabolism that describe the relationships between the metabolic rate, body size, and temperature in biological systems, ranging from the cellular to the ecosystem level. MST, which integrates the West, Brown, and Enquist network (WBE) model (West et al. 1997) and the ecological and evolutionary extensions (West et al. 1999), the metabolic theory of ecology (Brown et al. 2004), other existing network theories, and empirical knowledge, offers a unified framework to connect scaling phenomena mechanistically. The theory posits that metabolic rate scales with an organism's body mass to the 3/4 power. This relationship is thought to be a consequence of the fractal nature of resource distribution networks within organisms and the energetic and material constraints on biological processes. There has been considerable debate on how best to apply and test MST (see discussion and references in Price et al. 2012). Extension of the theory that relaxes some of the core assumptions of the theory can incorporate variation in biological scaling and can provide a basis for understanding the drivers of variation in biological scaling (Enquist and Bentley 2012, Savage et al. 2008).

Efforts to integrate organismal metabolic functions with ecosystem-based approaches have been used to estimate energy flux and storage from localized ecosystems to the biosphere (Hatton et al. 2015, Michaletz et al. 2014, Schramski et al. 2015). The use of metabolic scaling theory has been elaborated and applied to specific problems, including on water usage ranging from individual trees, to species, to forests and ecosystems, and has revealed that metabolic scaling varies based on complex trait interactions and covariance (Sperry et al. 2012)

Trait-Based Theory

A trait-based approach in ecology provides measures of the traits of individuals in a species, from genomes/genes to cells, and to the physiology of the whole organism, offering several advantages over traditional taxonomic (e.g., species) or traditional population-dynamic-based approaches (McGill et al. 2006). Trait-based approaches are relevant to CSB because they allow for broader ecological generalizations and can offer mechanistic insights into how and why particular patterns in diversity, abundance, or ecosystem function emerge. Because of the focus on processes and functions, a trait-based approach enables research to be easily incorporated into various models, including those for climate change predictions, land-use change, or invasion ecology. Last, because traits are linked to function, traits can be powerful predictors of how ecological communities will change in response to environmental fluctuations.

Trait Driver Theory: An Integration of Trait-Based
Theory and Metabolic Scaling Theory

Trait driver theory (TDT) provides the potential to predict biogeographic patterns and processes and to estimate past and potential future community responses to climatic changes. TDT is relevant to CSB because it facilitates a more mechanistic understanding of the effects of environmental drivers of change, such as drought and temperature, on functional diversity and variations in growth, mortality, and productivity. TDT serves as a framework for (i) synthesizing mechanistic theory within ecology; (ii) reframing the predictions of numerous ecological theories formerly built on species coexistence theory, in terms of trait distribution dynamics; and (iii) incorporating specific traits, particularly body size and carbon acquisition traits, to "scale up" and forge a link between ecosystem functioning and species assemblage dynamics across climatic gradients (Enquist et al. 2015). By offering a robust theoretical foundation, TDT applies to larger spatial and temporal scales that influence ecological and evolutionary processes, including ecosystem-level metabolic processes, such as productivity, turnover and carbon, and nutrient cycling.

Theory of Complex System Dynamics

The application of the theory of complex systems to temporal and spatial aspects of ecosystem functions is especially relevant to CSB science. Indicators such as increasing lag, autocorrelation, and variance provide generic early warning signals (EWSs) of the tipping point of ecosystems to a new state by detecting how dynamics slow down near the transition. EWSs of tipping points are vital to anticipate system collapse or other sudden shifts (Bury et al. 2021). Some of this work has begun to include coupled human–environment system models (Bauch et al. 2016). For example, ESMs integrate the interactions of atmosphere, ocean, land, ice, and biosphere to estimate the state of regional and global climate change under a wide variety of conditions. ESMs are distinguished from climate models by their ability to simulate the feedback from biology that impacts biosphere-level processes. They include roles of biology and feedbacks (i.e., vegetation functional types), but a challenge is that they depend on discrete functional types based on classifications.

New Theory Development and Synthesis

The committee recognizes the urgent need for deeper theoretical development in CSB, building upon established theory that facilitates integration across scales. While current theories provide a solid foundation, substantial gaps remain that must be addressed to enhance theory synthesis and achieve a more predictive and integrative understanding of biological systems from a systems perspective. The committee identifies four critical areas for advancement, aiming to refine and broaden the theoretical scope to more effectively interpret and utilize multiscale data.

First, expanding the scope of theoretical development is imperative to address several undertheorized areas in biology that are crucial for the resilience and adaptability of ecosystems and their organisms. This includes linking evolutionary processes to biogeochemical cycling, organismal acclimation, and adaptation, as well as understanding the implications for human health and zoonoses. For instance, deeper insights into evolutionary mechanisms could elucidate how species adapt to climate change, resist diseases, or manage geographical expansions. Similarly, advancing theories on biogeochemical cycling could clarify the interactions between nutrient flows and ecosystem functioning under anthropogenic stress. Furthermore, theories that explore organismal acclimation and adaptation are essential for predicting responses to rapid environmental changes, which are vital for formulating effective conservation strategies and sustainability plans. Integrating these theories with studies on human health and zoonoses can also bridge crucial knowledge gaps in how environmental changes promote the emergence and spread of diseases.

Second, there is a strong need to integrate and potentially unify existing ecological and evolutionary theories to create a more cohesive framework capable of explaining a broader range of biological phenomena, thereby enhancing the predictive power of CSB. Merging concepts such as niche theory with metapopulation dynamics could shed new light on species distributions under environmental stress, while incorporating evolutionary game theory could provide deeper insights into adaptive behaviors in fluctuating ecosystems. Such theoretical integration is crucial not only for solidifying the scientific foundations of CSB but also for fostering interdisciplinary collaborations that bring together diverse fields such as ecology, evolutionary biology, climatology, and public health.

Third, advancing CSB theory requires integration of biological feedbacks and human impacts. It is essential to address the complex interactions between biological and social systems, particularly how human activities influence continental-scale biological processes. CSB must account for the role of human populations in shaping continental-scale biological processes. By integrating CSB theory with established ecological and biological theories, we enhance continuity and leverage existing knowledge, highlighting how human activities influence large-scale ecological dynamics. Furthermore, theory development needs to consider conditions not only within a place but also interactions with other places nearby and far away (Frans and Liu 2024).

Fourth, to support this collaborative and cross-scale theory development, CSB could establish dedicated platforms or working groups aimed at synthesizing and advancing ecological and evolutionary theory. These groups could operate within existing organizational structures or through newly established interdisciplinary institutes designed to encourage cross-disciplinary collaborations. Promoting regular workshops, symposia, and joint research initiatives will be crucial to encourage dialogue and the exchange of ideas. Furthermore, integrating diverse knowledge systems, including indigenous and local ecological knowledge (Kimmerer and Artelle 2024), would provide valuable perspectives and enrich the theoretical frameworks, making them more applicable to real-world scenarios. This interdisciplinary and inclusive approach is essential for addressing the complex challenges posed by global ecological changes and ensuring the development of robust, applicable, and inclusive theory within CSB.

CHALLENGES IN CONNECTING RESEARCH ACROSS SCALES

Realizing the promise of CSB will require overcoming the challenge of connecting research across scales. The challenge is two-fold. **The *first challenge* is to integrate data obtained from very different methodologies across spatial and temporal scales.** Theory could help guide this integration by identifying which variables, scales, and processes are important. **The *second challenge* lies in the need for the development of theoretical frameworks to keep pace with the exponential rise in ecological and environmental data across spatial and temporal scales, which often have biases related to technology (e.g., 'omics are primarily done at comparatively small scales), geography (high-income countries dominate), and social factors (lack of capacity or financial resources to employ complex, expensive methodologies).** CSB, with its domain of inquiry ranging from the microscopic to the macroscopic of continental scale, is particularly challenged by data accrual outpacing the development of theory. In addition, analyses of biological collections and biological samples and observations are faced with numerous biases that can influence understanding and can skew or derail theoretical developments.

The Data Deluge and CSB: Theory as High Ground in the Flood of Big-Data for CSB

Some examples of major biases that need to be addressed in CSB theory follow.

Geographic and Temporal Sampling Bias

Despite major advances in global biodiversity, trait, and ecosystem data availability, trait data are disproportionately only available for the Global North, with major data shortfalls in biodiverse regions in the Global South. This geographic bias reflects colonial history, population density, ease of access, and numerous logistical challenges that inject bias and limit both fundamental and applied science (Feng et al. 2022, Maitner et al. 2023, Park et al. 2023, Schimel et al. 2015). Furthermore, this sampling bias limits attempts to analyze scale-dependent patterns and global processes. In addition, data that reflect temporal and spatial components of systems often are not available.

Availability Bias

Data are heterogeneous and often do not follow open science FAIR (findability, accessibility, interoperability, and reusability) principles, which include a core set of criteria used that enhance the ability to automatically find and use or reuse data (Wilkinson et al. 2016). The heterogeneous nature of the information available and lack of adherence to FAIR data principles for comparative analyses of datasets is a significant challenge (Gallagher et al. 2020). Data are scattered across publications and databases with variable formats, units, and methodologies. This fragmentation of data, often scarce metadata, and lack of temporal- and spatial-scale data impedes the synthesis and analysis of trait data, limiting the potential for cross-study comparisons, and broader insights in theory. In addition to FAIR principles, CARE (collective benefit, authority to control, responsibility, and ethics) principles are important to make sure that the governance and rights of the owners who provide knowledge of various environments and ecosystems that produce experimental datasets are included in discussions of results with consideration of rights and access to data (Carroll et al. 2021).

Improper Data Practices and Incomplete Representation of Functions

"As researchers increasingly utilize publicly available databases to guide research questions and conduct analyses, the abundance of currently available data will influence the trajectory of future research and data collection" (Augustine et al. 2024). Multiple issues associated with incorrect dataset uploading and inadequate curation of database data are suspected in increasing error (Augustine et al. 2024). There is also considerable bias and variation in how data are sampled. For example, in plants, most trait data are poorly sampled. In the TRY plant trait,[1] mean trait completeness is less than 1 percent, although recent efforts focused on particular traits (e.g., growth form) have reached high levels of completeness.

Biodiversity Bias

Biodiversity is a multidimensional construct, the three most common dimensions in use being taxonomic diversity (e.g., species richness), phylogenetic diversity (e.g., cumulative branch length of a phylogenetic tree encompassing species under investigation), and functional diversity (e.g., trait-based diversity metrics), although there are more (e.g., trophic, landscape, genomic, etc.) (Naeem et al. 2016). The majority of research in biodiversity across all scales is biased toward unidimensional research, with taxonomic diversity being the dominant dimension. Although studies of phylogenetic and functional diversity are on the rise, as is multidimensional biodiversity research, studies across all scales are deficient in their coverage of species. For example, plant species are not sampled representatively across the tree of life, with some clades (e.g., Poaceae) being relatively well sampled while others (e.g., Orchidaceae) are relatively poorly sampled.

Challenge of the "Siren Call" of Machine Learning and Artificial Intelligence

Navigating the expanding influence of big data, ML, and AI within the scientific community presents a substantial challenge. The rapid acceleration of science through data-derived modeling is undeniable (Krenn et al. 2022), yet it introduces a critical trade-off. Although ML and AI can deliver remarkable accuracy by leveraging existing datasets, their tendency to overfit and the lack of transparent, mechanistic models raise concerns (Mitchell 2019). These algorithms excel at generating short-term forecasts but

---

[1] See https://www.try-db.org/TryWeb/Home.php (accessed April 27, 2024).

often falter in more extended projections that divert focus from the pursuit of fundamental mechanistic insights in the study of complex adaptive systems.

The efficacy of integrating increased biological detail and mechanisms into forecasting models, without simultaneously broadening our theoretical foundations, remains uncertain. ML and data-derived modeling are frequently seen as sophisticated regression techniques (Mitchell 2019), sharing both their strengths and inherent limitations. Conversely, coarse-grained methods from theory offer the potential to generate reliable predictions beyond the immediate scope of training data, uncover novel simplifications, and mitigate the risk of overfitting. While regression-based methods, including ML, have their place in scenarios where theoretical underpinnings are underdeveloped or for in-sample predictions, there is much potential synergy between theory-driven and data-driven discovery. AI and ML methods should be seen as tools to do data-driven discovery more quickly. Further, the development of innovative coarse-grained ML techniques holds promise (e.g., Brunton et al. 2016, Han et al. 2018, Schmidt and Lipson 2009, Udrescu and Tegmark 2020). These emerging approaches could bridge the gap between data-driven discovery, accuracy, and the quest for mechanistic understanding, underscoring the need for a balanced integration of theory, ML, and AI in advancing the science of complex ecological systems (Han et al. 2023).

Challenges to Developing Theory Across Scales (Molecules to Biosphere) and Implementation of Theory for Informing Hypotheses and Experimentation Across Scales

Challenge of Trade-Offs

The development of theory across scales, from molecules to the biosphere, presents significant challenges, particularly in modeling and applying theory to inform hypotheses and experimentation. One major hurdle is the inherent trade-offs involved in scientific modeling, as identified half a century ago by Levins (1966). These trade-offs often manifest in the choice between modeling a few locations or species in great detail versus many locations and species more superficially. An overly abstract or reductionistic representation may inadequately highlight pattern and identify the essence of the system. For CSB-related experimentation, this may translate as balancing model accuracy with generalizability, essential for applying theory across varied ecosystems and scales. Further, a representation of exact replication becomes cumbersome and essentially indistinguishable from the entity it seeks to describe (Enquist et al. 2024). Balancing between virtues such as accuracy, precision, realism, and generality becomes an inherent part of the endeavor of theory development. Accepting such trade-offs is not a sign of theoretical weakness. Instead, trade-offs are indicative of the nuanced choices scientists make in their journey to understand complex phenomena.

Another trade-off is navigating the integration of big data, ML, and statistical complexity. Data-driven models, while accelerating scientific progress (Krenn et al. 2022) often produce accurate but potentially overfitted predictions that lack transparency in mechanisms (Mitchell 2019, West 2013). This limitation hampers long-term forecasting and may shift the focus away from developing deep mechanistic insights essential for understanding complex adaptive systems (Enquist et al. 2024). As we develop theory across scales—from molecules to the biosphere—it is crucial to maintain a balance between utilizing advanced computational tools and fostering robust theoretical frameworks. This approach ensures that our reliance on modern technologies complements rather than supplants the pursuit of comprehensive, mechanistically informed scientific theory, thereby enhancing both immediate data analysis and long-term predictive capabilities.

Challenge of Biodiversity Complexity

Biodiversity, as is commonly defined, includes all dimensions, from ecological to evolutionary, across all dimensions of space and time, but in practice, theoretical and empirical approaches focus on single dimensions within scales. An example of the challenge of trade-offs in the inherent tension between modeling a select number of species or locations or a specific biological process in depth versus attempting to capture a broad array of species and functions by working at higher orders, such as tree-species richness and Normalized Difference Vegetation Index across kilometers. Addressing these trade-offs, especially when constructing process-based models for numerous locations or species, among others, remains one of biodiversity science's most formidable challenges (Figure 3-2).

The committee posits that these challenges can be effectively navigated, irrespective of the chosen study group or modeling framework, by leveraging modeling techniques that harness strength across locations and dimensions of biodiversity, by filling data voids using proxies, amalgamating varied data sources, and doing so across different scales. Opportunities exist to further develop existing theory to help guide process-based models for many sites and across multiple levels of biodiversity by using hierarchical and inverse modeling methods, to fill data gaps, integrate diverse datasets, and model across biological and spatial scales (Evans 2016, Levin 2000).

CONCLUSIONS ON DEVELOPING THEORY TO CONNECT RESEARCH ACROSS SCALES

Theoretical considerations should guide the collection and management of big data, that is, what data that we should collect and what we need to collect. In addition, theory should support our understanding of the causes and consequences of continental-scale biodiversity and resilience in the context of major global change (Figure 3-2). The committee concluded that theory is especially needed in the following three areas.

Conclusion 3-1: *Theory is needed that links research at multiple organizational, spatial, and temporal scales, from the micro to meter to landscapes up to the biosphere.* The multidimensional and hierarchical multiscale nature of biodiversity requires solutions that can address cross-scale questions and identify cross-scale phenomena (Isbell et al. 2017, Soranno et al. 2014). For this approach, theory is needed that meshes with our current technologies and informatics that collectively monitor biosphere processes. Consistent with the theme in Chapter 2 about the need for integrated yet flexible frameworks for CSB, theory also needs to be based on conceptual frameworks that integrate multiscale data. These include molecular, microbial processes, genomes, environmental DNA, metagenomics, metatranscriptomics, stable isotope labeling, and metabolomics. These data sources are crucial for linking local ecological and physiological processes of organisms to broader patterns and data collection efforts such as the distribution of species, movement of individuals and species, the functioning of ecosystems, and the flux of material and matter through the biosphere at multiple scales. This integration will enable the refinement and development of CSB theory, enhancing our ability to model and manage environmental changes effectively. By incorporating larger-scale data from remote sensing, tower-based systems, global animal tracking, and sensor networks, we can enrich this framework, providing a more comprehensive understanding necessary for predictive modeling and sustainable ecosystem management.

Conclusion 3-2: Theory is needed to improve climate and global change models by including biological feedbacks. Biological processes that result in feedbacks to ecosystems and climate are a challenge to incorporate into climate and global change theories, presenting considerable uncertainty (Figure 3-2). The inclusion of biological feedback to continental-scale models of global change will enhance our ability to predict future trends and identify cross-scale solutions and will be a key component of clarifying and improving climate and global change models. Refinement of biological feedback theories into continental-scale models and extension to climate and global change theories will improve our ability to both predict future trends as well as identify solutions that cut across scales.

Conclusion 3-3: Theory is needed that incorporates the effects of human-induced environmental changes (including climate change) to predict changes within an ecosystem and to assess metacoupled cascading effects across adjacent and distant systems. Theory is needed to predict interactions among system components across all scales that impact adjacent and distant environments. The inclusion of theory that incorporates human activity will enable the prediction of synergistic, cascading, or trade-off effects on resilience and sustainability of ecosystems and the biosphere across time and space.

To summarize, theory is important to CSB because a well-defined theory with specific testable predictions that can be proven wrong by confrontation with data can provide important insights for moving a field in the right direction (Enquist et al. 2024). Core attributes of successful theories for CSB are that they need to be applicable at various biological organizational, spatial, and temporal scales; they need to be able to inform the collection and management of big data and ensure that data collection aligns with the needs of understanding large-scale biological and ecological processes; they need to unify disparate functional processes from microbial to ecological and physiological processes, including material flux through the biosphere; and they need to be transparent about the role of different perspectives, experiences, and knowledge foundations of research team members in driving innovation and analyses.

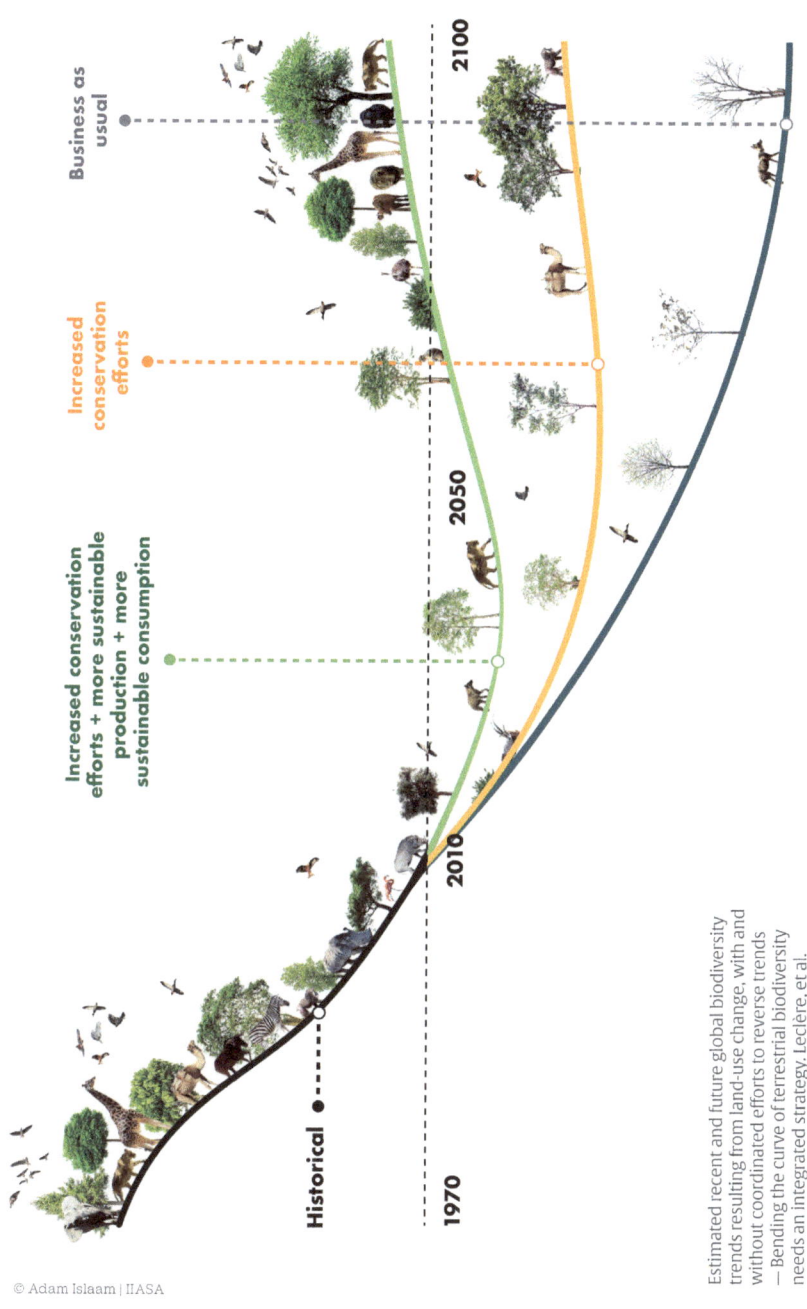

Estimated recent and future global biodiversity trends resulting from land-use change, with and without coordinated efforts to reverse trends — Bending the curve of terrestrial biodiversity needs an integrated strategy. Leclère, et al.

FIGURE 3-2 Hypothetical biodiversity curves illustrating different scenarios and impacts on biodiversity targets. The figure depicts the feasibility of reversing global biodiversity decline while balancing food provision and other land uses.
SOURCE: Adapted from Leclère et al. (2020), https://sevenseasmedia.org/bending-the-curve-of-biodiversity-loss/.

The committee identified a number of challenges in applying existing theories and developing new theories to support CSB. Moving from data to insight is challenged by the unprecedented levels of detail that can be observed at any scale. Technologies to collect data are far outpacing the derivation of theoretical frameworks that can process this information across multiple scales and glean understanding and predictive ability. Directly related is the expanding influence of big data, ML, and AI. Building on these are the inherent trade-offs involved in scientific modeling, for example, in the choice between modeling a few locations or species in great detail versus many locations and species more superficially. Balancing accuracy, precision, realism, and generality is a challenge inherent to all biological theory development, including that supporting CSB.

Further, theory needs to address potential sources of bias, for example, geographic and temporal sampling bias, availability bias, incomplete representation of functions, and biodiversity bias. A final challenge is the complexity of life itself—as noted previously, biodiversity includes scales from ecological to evolutionary and all dimensions of space and time. The committee is confident that, supported by the tools, networks, and training described in Chapters 4 and 5, these challenges can be effectively navigated.

REFERENCES

Abrams, P.A. 1986. Character displacement and niche shift analyzed using consumer-resource models of competition. *Theoretical Population Biology* 29:107-160. https://doi.org/10.1016/0040-5809(86)90007-9.

Arias, P.A., N. Bellouin, E. Coppola, R.G. Jones, G. Krinner, J. Marotzke, V. Naik, M.D. Palmer, G.-K. Plattner, J. Rogelj, M. Rojas, J. Sillmann, T. Storelvmo, P.W. Thorne, B. Trewin, et al., 2021: Technical summary. Pp. 33-144 In *Climate Change 2021: The Physical Science Basis. Contribution of Working Group I to the Sixth Assessment Report of the Intergovernmental Panel on Climate Change*, V. Masson-Delmotte, P. Zhai, A. Pirani, S.L. Connors, C. Péan, et al., eds., Cambridge, UK, and New York: Cambridge University Press.

Arora, V.K., A. Katavouta, R.G. Williams, C.D. Jones, V. Brovkin, P. Friedlingstein, J. Schwinger, et al. 2020. Carbon-concentration and carbon-climate feedbacks in CMIP6 models, and their comparison to CMIP5 models. *Biogeosciences* 17:4173-4222. https://doi.org/10.5194/bg-17-4173-2020.

Augustine, S.P., I. Bailey-Marren, K.T. Charton, N.G. Kiel, and M.S. Peyton. 2024. Improper data practices erode the quality of global ecological databases and impede the progress of ecological research. *Global Change Biology* 30(1):e17116. https://doi.org/10.1111/gcb.17116.

Bauch, C.T., R. Sigdel, J. Pharaon, and M. Anand. 2016. Early warning signals of regime shifts in coupled human–environment systems. *Proceedings of the National Academy of Sciences of the United States of America* 113:14560-14567 https://doi.org/10.1073/pnas.1604978113.

Bennett, E.M., M. Solan, R. Biggs, T. McPhearson, A.V Norström, P. Olsson, L. Pereira, G.D. Peterson, C. Raudsepp-Hearne, et al. 2016. Bright spots: Seeds of a good Anthropocene. *Frontiers in Ecology and the Environment* 14:441-448. https://doi.org/10.1002/fee.1309.

Brown, J.H., J.F. Gillooly, A.P. Allen, V.M. Savage, and G.B. West. 2004. Toward a metabolic theory of ecology. *Ecology* 85:1771-1789. https://doi.org/10.1890/03-9000.

Brunton, S.L., J.L. Proctor, and J.N. Kutz. 2016. Discovering governing equations from data by sparse identification of nonlinear dynamical systems. *Proceedings of the National Academy of Sciences of the United States of America* 113:3932-3937. http://dx.doi.org/10.1073/pnas.1517384113.

Bury, T.M., R.I. Sujith, I. Pavithran, and C.T. Bauch. 2021. Deep learning for early warning signals of tipping points. *Proceedings of the National Academy of Sciences of the United States of America* 118(39): e2106140118 https://doi.org/10.1073/pnas.2106140118.

Cadotte, M.W., J. Cavender-Bares, D. Tilman, and T.H. Oakley. 2009. Using phylogenetic, functional and trait diversity to understand patterns of plant community productivity. *PLoS ONE* 4:e5695. https://doi.org/10.1371/journal.pone.0005695.

Carroll, S.R., E. Herczog, M. Hudson, K. Russell, and S. Stall. 2021. Operationalizing the CARE and FAIR principles for Indigenous data futures. *Scientific Data* 8:108. https://doi.org/10.1038/s41597-021-00892-0

Ceballos, G., and P. R. Ehrlich. 2018. The misunderstood sixth mass extinction. *Science* 360: 1080-1081. https://doi.org/10.1126/science.aau0191.

Chaparro-Pedraza, P.C., and A. M. de Roos. 2020. Ecological changes with minor effect initiate evolution to delayed regime shifts. *Nature Ecology & Evolution* 4:412-418. https://doi.org/10.1038/s41559-020-1110-0.

Collatz, G.J., J.T. Ball, C. Grivet, and J.A. Berry. 1991. Physiological and environmental regulation of stomatal conductance, photosynthesis and transpiration: A model that includes a laminar boundary layer. *Agricultural and Forest Meteorology* 54:107-136. https://doi.org/10.1016/0168-1923(91)90002-8.

Coveney, P.V., E.R. Dougherty, and R.R. Highfield. 2016. Big data need big theories too. *Philosophical Transactions of the Royal Society A: Mathematical, Physical and Engineering Sciences* 374(2080): 20160153. https://doi.org/10.1098/rsta.2016.0153.

Díaz, S., J. Settele, E. Brondizio, H.T. Ngo, J. Agard, A. Arneth, P. Balvanera, et al. 2019. Pervasive human-driven decline of life on Earth points to the need for transformative change. *Science* 366(6471). https://doi.org/10.1126/science.aax3100.

Dietze, M.C., A. Fox, L.M. Beck-Johnson, J.L. Betancourt, M.B. Hooten, C.S. Jarnevich, T.H. Keitt, et al. 2018. Iterative near-term ecological forecasting: Needs, opportunities, and challenges. *Proceedings of the National Academy of Sciences of the United States of America* 115:1424-1432. https://doi.org/10.1073/pnas.1710231115.

Ditlevsen, P.D., and S.J. Johnsen. 2010. Tipping points: Early warning and wishful thinking. *Geophysical Research Letters* 37(19):L19703. https://doi.org/10.1029/2010GL044486.

Doak, D.F., J.A. Estes, B.S. Halpern, U. Jacob, D.R. Lindberg, J. Lovvorn, D.H. Monson, et al. 2008. Understanding and predicting ecological dynamics: Are major surprises inevitable? *Ecology* 89:952-961. https://doi.org/10.1890/07-0965.1.

Dornelas, M., N.J. Gotelli, B. McGill, H. Shimadzu, F. Moyes, C. Sievers, and A.E. Magurran. 2014. Assemblage time series reveal biodiversity change but not systematic loss. *Science* 344: 296-299. https://doi.org/10.1126/science.1248484.

Drake, J.M., S.M. O'Regan, V. Dakos, S. Kéfi, and P. Rohani. 2020. Alternative stable states, tipping points, and early warning signals of ecological transitions. Pp. 263-284 in *Theoretical Ecology: Concepts and Applications*, K.S. McCann and G. Gellner, eds. Oxford University Press.

Dudney, J., and K.N. Suding. 2020. The elusive search for tipping points. *Nature Ecology & Evolution* 4:1449–1450. https://doi.org/10.1038/s41559-020-1273-8.

Emanuel, K. 2020. The relevance of theory for contemporary research in atmospheres, oceans, and climate. *AGU Advances* 1(2):e2019AV000129. https://doi.org/10.1029/2019AV000129.

Enquist, B.J., and L.P. Bentley. 2012. Land plants: New theoretical directions and empirical prospects. Pp. 164-187 in *Metabolic Ecology: A Scaling Approach*, R. M. Sibly, J. H. Brown, and A. Kodric-Brown, eds. John Wiley & Sons.

Enquist, B.J., J. Norberg, S.P. Bonser, C. Violle, C.T. Webb, A. Henderson, L.L. Sloat, and V.M. Savage. 2015. Scaling from traits to ecosystems: Developing a general trait driver theory via integrating trait-based and metabolic scaling theory. *Advances in Ecological Research* 52:249-318. https://doi.org/10.1016/bs.aecr.2015.02.001.

Enquist, B.J., C.P. Kempes, and G.B. West. 2024. Developing a predictive science of the biosphere requires the integration of scientific cultures. *Proceedings of the National Academy of Sciences of the United States of America* 121(19):e2209196121. https://doi.org/10.1073/pnas.2209196121.

Evans, E.W. 2016. Biodiversity, ecosystem functioning, and classical biological control. *Applied Entomology and Zoology* 51:173-184. https://doi.org/10.1007/s13355-016-0401-z.

Farquhar, G.D., S. von Caemmerer, and J.A. Berry. 1980. A biochemical model of photosynthetic CO_2 assimilation in leaves of C3 species. *Planta* 149: 78-90. https://doi.org/10.1007/BF00386231.

Feng, X., B.J. Enquist, D.S. Park, B. Boyle, D.D. Breshears, R.V. Gallagher, A. Lien, E.A. Newman, J.R. Burger, B.S. Maitner, C. Merow, Y. Li, K.M. Huynh, K. Ernst, et al. 2022. A review of the heterogeneous landscape of biodiversity databases: Opportunities and challenges for a synthesized biodiversity knowledge base. *Global Ecology and Biogeography* 31(7):1242-1260. https://doi.org/10.1111/geb.13497.

Fisher, R.A. C.D. Koven, W.R.L. Anderegg, B.O. Christoffersen, M.C. Dietze, C.E. Farrior, J.A. Holm, et al. 2018. Vegetation demographics in Earth system models: A review of progress and priorities. *Global Change Biology* 24:35-54. https://doi.org/10.1111/gcb.13910.

Folke, C., Å. Jansson, J. Rockström, P. Olsson, S.R. Carpenter, F.S. Chapin III, A.-S. Crépin, et al. 2011. Reconnecting to the biosphere. *AMBIO* 40:719-738. https://doi.org/10.1007/s13280-011-0184-y.

Franklin, O., S.P. Harrison, R. Dewar, C.E. Farrior, A. Brännström, U. Dieckmann, S. Pietsch, D. Falster, W. Cramer, M. Loreau, and H. Wang. 2020. Organizing principles for vegetation dynamics. *Nature Plants* 6: 444-453. https://doi.org/10.1038/s41477-020-0655-x.

Frans, V.F., and J. Liu. 2024. Gaps and opportunities in modelling human influence on species distributions in the Anthropocene. *Nature Ecology & Evolution* 8:1365–1377. https://doi.org/10.1038/s41559-024-02435-3.

Friedlingstein, P., P. Cox, R. Betts, L. Bopp, W. von Bloh, V. Brovkin, P. Cadule, S. Doney, M. Eby, I. Fung, et al. 2006. Climate–carbon cycle feedback analysis: Results from the C4MIP model intercomparison. *Journal of Climate* 19:33373353. https://doi.org/10.1175/JCLI3800.1.

Friedlingstein, P., M. Meinshausen, V.K. Arora, C.D. Jones, A. Anav, S.K. Liddicoat, and R. Knutti. 2014. Uncertainties in CMIP5 climate projections due to carbon cycle feedbacks. *Journal of Climate* 27:511-526. https://doi.org/10.1175/JCLI-D-12-00579.1.

Gallagher, R.V., D.S. Falster, B.S. Maitner, R. Salguero-Gómez, V. Vandvik, W.D. Pearse, F.D. Schneider, J. Kattge, J.H. Poelen, J.S. Madin, M.J. Ankenbrand, C. Penone, X. Feng, et al. 2020. Open Science principles for accelerating trait-based science across the tree of life. *Nature Ecology & Evolution* 4(3):294-303. https://doi.org/10.1038/s41559-020-1109-6.

Graven, H., R.F. Keeling, S.C. Piper, P.K. Patra, B.B. Stephens, S.C. Wofsy, L.R. Welp, C. Sweeney, P.P. Tans, J.J. Kelley, B.C. Daube, E.A. Kort, G.W. Santoni, and J.D. Bent. 2013. Enhanced seasonal exchange of CO_2 by northern ecosystems since 1960. *Science* 341:1085-1089. https://doi.org/10.1126/science.1239207.

Guisan, A., and W. Thuiller. 2005. Predicting species distribution: offering more than simple habitat models. *Ecology Letters* 8:993-1009. https://doi.org/10.1111/j.1461-0248.2005.00792.x.

Guisan, A., and N.E. Zimmermann. 2000. Predictive habitat distribution models in ecology. *Ecological Modelling* 135:147-186. https://doi.org/10.1016/S0304-3800(00)00354-9.

Haldane, J.B.S. 1927. Science and theology as art-forms. *Possible Worlds* 227.
Hampton, S.E., C.A. Strasser, J.J. Tewksbury, W.K. Gram, A.E. Budden, A.L. Batcheller, C.S. Duke, and J.H. Porter. 2013. Big data and the future of ecology. *Frontiers in Ecology and the Environment* 11:156-162 https://doi.org/10.1890/120103.
Han, J., A. Jentzen, and E. Weinan. 2018. Solving high-dimensional partial differential equations using deep learning. *Proceedings of the National Academy of Sciences of the United States of America* 115:85058510. https://doi.org/10.1073/pnas.171894211.
Han, B.A., K.R. Varshney, S. LaDeau, A. Subramaniamc, K.C. Weathers, and J. Zwart. 2023. A synergistic future for AI and ecology. *Proceedings of the National Academy of Sciences of the United States of America* 120(38):e2220283120. https://doi.org/10.1073/pnas.2220283120.
Harrison, S.P., W. Cramer, O. Franklin, I.C. Prentice, H. Wang, A. Brännström, H. De Boer, U. Dieckmann, J. Joshi, T.F. Keenan, and A. Lavergne. 2021. Eco-evolutionary optimality as a means to improve vegetation and land-surface models. *New Phytologist* 231:2125-2141. https://doi.org/10.1111/nph.17558.
Hatton, I.A., K.S. McCann, J.M. Fryxell, T.J. Davies, M. Smerlak, A.R. Sinclair, and M. Loreau. 2015. The predator-prey power law: Biomass scaling across terrestrial and aquatic biomes. *Science* 349(6252):aac6284. https://doi.org/10.1126/science.aac6284.
Hausfather, Z., H.F. Drake, T. Abbott, and G.A. Schmidt. 2020. Evaluating the performance of past climate model projections. *Geophysical Research Letters* 47:e2019GL085378. https://doi.org/10.1029/2019GL085378.
Hubbell, S.P. 2001. *The Unified Neutral Theory of Biodiversity and Biogeography.* Princeton University Press.
Huntzinger, D.N., A.M. Michalak, C. Schwalm, P. Ciais, A.W. King, F.Y. Fang, K. Schaefer, Y. Wei, et al. 2017. Uncertainty in the response of terrestrial carbon sink to environmental drivers undermines carbon-climate feedback predictions. *Scientific Reports* 7:4765. https://doi.org/10.1038/s41598-017-03818-2.
IPCC (Intergovernmental Panel on Climate Change). 2014. Carbon and other biogeochemical cycles. Pp. 465-570 in *Climate Change 2013—The Physical Science Basis: Working Group I Contribution to the Fifth Assessment Report of the Intergovernmental Panel on Climate Change.* Cambridge University Press.
Isbell, F., A. Gonzalez, M. Loreau, J. Cowles, S. Díaz, A. Hector, G.M. Mace, D.A. Wardle, M.I. O'Connor, J.E. Duffy, L.A. Turnbull, P.L. Thompson, and A. Larigauderie. 2017. Linking the influence and dependence of people on biodiversity across scales. *Nature* 546:65-72. https://doi.org/10.1038/nature22899.
Jung, M., A. Arnell, X. de Lamo, S. García-Rangel, M. Lewis, J. Mark, C. Merow, L. Miles, I. Ondo, S. Pironon, C. Ravilious, M. Rivers, D. Schepaschenko, O. Tallowin, A. van Soesbergen, R. Govaerts, et al. 2021. Areas of global importance for conserving terrestrial biodiversity, carbon and water. *Nature Ecology & Evolution* 5:1499-1509. https://doi.org/10.1038/s41559-021-01528-7.
Keith, D.W. 2021. Toward constructive disagreement about geoengineering. *Science* 374:812-815. https://doi.org/10.1126/science.abj1587.
Kimmerer, R.W., and K.A. Artelle. 2024. Time to support Indigenous science. *Science* 383:243. https//doi.org/10.1126/science.ado0684.
Kivlin, S.N., C.V. Hawkes, M. Papeş, K.K. Treseder, and C. Averill. 2021. The future of microbial ecological niche theory and modeling. *New Phytologist* 231:508-511. https://doi.org/10.1111/nph.17373.
Krenn, M., R. Pollice, S. Y. Guo, M. Aldeghi, A. Cervera-Lierta, P. Friederich, G. dos Passos Gomes, F. Häse, A. Jinich, A.K. Nigam, Z. Yao, and A. Aspuru-Guzik. 2022. On scientific understanding with artificial intelligence. *Nature Reviews Physics* 4:761-769. https://doi.org/10.1038/s42254-022-00518-3.

Leclère, D., M. Obersteiner, M. Barrett, S.H.M. Butchart, A. Chaudhary, A. De Palma, F.A.J. DeClerck, et al. 2020. Bending the curve of terrestrial biodiversity needs an integrated strategy. *Nature* 585:551-556. https://doi.org/10.1038/s41586-020-2705-y.

Lenton, T.M. 2013. Environmental tipping points. *Annual Review of Environment and Resources* 38:1-29. https://doi.org/10.1146/annurev-environ-102511-084654.

Levin, S.A., 2000. Multiple scales and the maintenance of biodiversity. *Ecosystems* 3:498-506. https://doi.org/10.1007/s100210000044.

Levins, R. 1966. The strategy of model building in population ecology. *American Scientist* 54:421-431.

Li, L., and Z. Ma. 2016. Testing the neutral theory of biodiversity with human microbiome datasets. *Scientific Reports* 6:31448. https://doi.org/10.1038/srep31448.

Leibold, M.A. 1995. The niche concept revisited: Mechanistic models and community context. *Ecology* 76:1371-1382. https://doi.org/10.2307/1938141.

Letten, A.D., P.-J. Ke, and T. Fukami. 2017. Linking modern coexistence theory and contemporary niche theory. *Ecological Monographs* 87:161-177. https://doi.org/10.1002/ecm.1242.

Maitner, B., R. Gallagher, J.C. Svenning, M. Tietje, E.H. Wenk, and W.L. Eiserhardt. 2023. A global assessment of the Raunkiæran shortfall in plants: Geographic biases in our knowledge of plant traits. *New Phytologist* 240(4):345-1354. https://doi.org/10.1111/nph.18999.

Malhi, Y., J. Franklin, N. Seddon, M. Solan, M.G. Turner, C.B. Field, and N. Knowlton. 2020. Climate change and ecosystems: Threats, opportunities and solutions. *Philosophical Transactions of the Royal Society B: Biological Sciences* 375(1794):20190104. http://doi.org/10.1098/rstb.2019.0104.

Marquet, P.A., A.P. Allen, J.H. Brown, J.A. Dunne, B.J. Enquist, J.F. Gillooly, P.A. Gowaty, J.L. Green, J. Harte, S.P. Hubbell, J. O'Dwyer, J.G. Okie, A. Ostling, M. Ritchie, D. Storch, and G.B. West. 2014. On theory in ecology. *Bioscience* 64(8):701-710. https://doi.org/10.1093/biosci/biu098.

McGill, B.J., B.J. Enquist, E. Weiher, and M. Westoby. 2006. Rebuilding community ecology from functional traits. *Trends in Ecology & Evolution* 21:178-185. https://doi.org/10.1016/j.tree.2006.02.002.

McGill, B.J., M. Dornelas, N.J. Gotelli, and A.E. Magurran. 2015. Fifteen forms of biodiversity trend in the Anthropocene. *Trends in Ecology & Evolution* 30:104-113. https://doi.org/10.1016/j.tree.2014.11.006.

McMichael, A.J. 2013. Globalization, climate change, and human health. *New England Journal of Medicine* 369:1335-1343. https://doi.org/10.1056/NEJMra1109341.

Michaletz, S.T., D. Cheng, A.J. Kerkhoff, and B.J. Enquist. 2014. Convergence of terrestrial plant production across global climate gradients. *Nature* 512: 39-43. https://doi.org/10.1038/nature13470.

Mitchell, M. 2019. *Artificial Intelligence: A Guide for Thinking Humans*. Farrar, Straus and Giroux.

Molotoks, A., R. Henry, E. Stehfest, J. Doelman, P. Havlik, T. Krisztin, P. Alexander, T.P. Dawson, and P. Smith. 2020. Comparing the impact of future cropland expansion on global biodiversity and carbon storage across models and scenarios. *Philosophical Transactions of the Royal Society B: Biological Sciences* 375:20190189 http://doi.org/10.1098/rstb.2019.0189.

Moorcroft, P.R. 2006. How close are we to a predictive science of the biosphere? *Trends in Ecology & Evolution* 21:400-407. https://doi.org/10.1016/j.tree.2006.04.009.

Morecroft, M.D., S. Duffield, M. Harley, J.W. Pearce-Higgins, N. Stevens, O. Watts, and J. Whitaker. 2019. Measuring the success of climate change adaptation and mitigation in terrestrial ecosystems. *Science* 366(6471):eaaw9256. https://doi.org/10.1126/science.aaw9256.

Morin, X., and M.J. Lechowicz. 2008. Contemporary perspectives on the niche that can improve models of species range shifts under climate change. *Biology Letters* 4:573-576. https://doi.org/10.1098/rsbl.2008.0181.

Naeem, S., J.E. Duffy, and E. Zavaleta. 2012. The functions of biological diversity in an age of extinction. *Science* 336:1401-1406. https://doi.org/10.1126/science.1215855.

Naeem, S., C. Prager, B. Weeks, A. Varga, D.F.B. Flynn, K. Griffin, R. Muscarella, M. Palmer, S. Wood, and W. Schuster. 2016. Biodiversity as a multidimensional construct: A review, framework and case study of herbivory's impact on plant biodiversity. *Philosophical Transactions of the Royal Society B: Biological Sciences* 283(1844):20153005. https://doi.org/10.1098/rspb.2015.3005.

Novick, K.A., T.F. Keenan, W.R.L. Anderegg, C.P. Normile, B.R.K. Runkle, E.E. Oldfield, G. Shrestha, et al. 2024. We need a solid scientific basis for nature- based climate solutions in the United States. *Proceedings of the National Academy of Sciences of the United States of America* 121:e2318505121. https://doi.org/10.1073/pnas.2318505121.

NRC (National Research Council). 2008. *The Role of Theory in Advancing 21st-Century Biology: Catalyzing Transformative Research*. Washington, DC: The National Academies Press. https://doi.org/10.17226/12026.

Park, D.S., X. Feng, S. Akiyama, M. Ardiyani, N. Avendaño, Z. Barina, B. Bärtschi, et al. 2023. The colonial legacy of herbaria. *Nature Human Behaviour* 7:1059-1068. https://doi.org/10.1038/s41562-023-01616-7.

Peters, W., I.R. van der Velde, E. van Schaik, J.B. Miller, P. Ciais, H.F. Duarte, I.T. van der Laan-Luijkx, M.K. van der Molen, M. Scholze, K. Schaefer, P.L. Vidale, A. Verhoef, D. Wårlind, D. Zhu, P.P. Tans, B., Vaughn, and J.W.C. White. 2018. Increased water-use efficiency and reduced CO_2 uptake by plants during droughts at a continental scale. *Nature Geosciences* 11:744-748.

Pillet, M., B. Goettsch, B.C. Merow, B. Maitner, X. Feng, P.R. Roehrdanz, and B.J. Enquist. 2022. Elevated extinction risk of cacti under climate change. *Nature Plants* 8:366-372. https://doi.org/10.1038/s41477-022-01130-0.

Potochnik, A. 2020. *Idealization and the Aims of Science*. The University of Chicago Press.

Price, C.A., J.S. Weitz, V.M. Savage, J. Stegen, A. Clarke, D.A. Coomes, P.S. Dodds, et al. 2012. Testing the metabolic theory of ecology. *Ecology Letters* 15(12):1465-1474. https://doi.org/10.1111/j.1461-0248.2012.01860.x.

Qiao, S., H. Wang, I.C. Prentice, and S.P. Harrison. 2020. Extending a first-principles primary production model to predict wheat yields. *Agricultural and Forest Meteorology* 287:107932. https://doi.org/10.1016/j.agrformet.2020.107932.

Qiao, S., H. Wang, I.C. Prentice, and S.P. Harrison. 2021. Optimality based modelling of climate impacts on global potential wheat yield. *Environmental Research Letters* 16(11):114013. https://doi.org/10.1088/1748-9326/ac2e38.

Rosindell, J., S.P. Hubbell, F. He, L.J. Harmon, and R.S. Etienne. 2012. The case for ecological neutral theory. *Trends in Ecology & Evolution* 27:203-208. https://doi.org/10.1016/j.tree.2012.01.004.

Rothman, D.H. 2017. Thresholds of catastrophe in the Earth system. *Science Advances* 3:e1700906. https://doi.org/10.1126/sciadv.1700906.

Ruckelshaus, M.H., S.T. Jackson, H.A. Mooney, K.L. Jacobs, K.S. Kassam, M.T.K. Arroyo, A. Báldi, et al. 2020. The IPBES Global Assessment: Pathways to action. *Trends in Ecology & Evolution* 35:407–414. https://doi.org/10.1016/j.tree.2020.01.009.

Savage, V.M., E.J. Deeds, and W. Fontana. 2008. Sizing up allometric scaling theory. *PLoS Computational Biology* 4(9):e1000171. https://doi.org/10.1371/journal.pcbi.1000171.

Scheiner, S.M., and M.R. Willig (eds.). 2011. *The Theory of Ecology*. University of Chicago Press.

Schimel, D., R. Pavlick, J.B. Fisher, G.P. Asner, S. Saatchi, P. Townsend, C. Miller, C. Frankenberg, K. Hibbard, and P. Cox. 2015. Observing terrestrial ecosystems and the carbon cycle from space. *Global Change Biology* 21:1762-1776. https://doi.org/10.1111/gcb.12822.

Schmidt, M., and H. Lipson. 2009. Distilling free-form natural laws from experimental data. *Science* 324:81-85. https://doi.org/10.1126/science.1165893.

Schramski, J.R., A.I. Dell, J.M. Grady, and J.H. Brown. 2015. Metabolic theory predicts whole-ecosystem properties. *Proceedings of the National Academy of Sciences of the United States of America* 112:2617-2622. https://doi.org/10.1073/pnas.1423502112.

Servedio, M.R., Y. Brandvain, S. Dhole, C.L. Fitzpatrick, E, E. Goldberg, C.A. Stern, J. Van Cleve, and D.J. Yeh. 2014. Not just a theory—The utility of mathematical models in evolutionary biology. *PLoS Biology* 12(12):e1002017. https://psycnet.apa.org/doi/10.1371/journal.pbio.1002017.

Smith, N.G., T.F. Keenan, I.C. Prentice, H. Wang, I.J. Wright, U. Niinemets, Y. Crous, et al. 2019. Global photosynthetic capacity is optimized to the environment. *Ecology Letters* 22:506-517.

Soranno, P.A., K.S. Cheruvelil, and E.G. Bissell. 2014. Cross-scale interactions: Quantifying multi-scaled cause–effect relationships in macrosystems. *Frontiers in Ecology and the Environment* 12:65-73. https://doi.org/10.1890/120366.

Sperry, J.S., D.D. Smith, V.M. Savage, B.J. Enquist, K.A. McCulloh, P.B. Reich, L.P. Bentley, and E.I. von Allmen. 2012. A species-level model for metabolic scaling in trees I. Exploring boundaries to scaling space within and across species. *Functional Ecology* 26:1054-1065. https://doi.org/10.1111/j.1365-2435.2012.02022.x.

Thomas, R.B., I.C. Prentice, H. Graven, P. Ciais, J.B. Fisher, M. Huang, D.N. Huntzinger, et al., 2016. Increased light-use efficiency in northern terrestrial ecosystems indicated by CO_2 and greening observations. *Geophysical Research Letters* 43:11339-11349. https://doi.org/10.1002/2016GL070710.

Thuiller, W., C. Albert, M.B. Araújo, P.M. Berry, M. Cabeza, A. Guisan, T. Hickler, G. Midgley, J. Paterson, F. Schurr, M.T. Sykes, and N. Zimmermann. 2008. Predicting global change impacts on plant species' distributions: Future challenges. *Perspectives in Plant Ecology, Evolution and Systematics* 9(3-4):137-152. https://doi.org/10.1016/j.ppees.2007.09.004.

Turner, M.G., W.J. Calder, G.S. Cumming, T.P. Hughes, A. Jentsch, S.L. LaDeau, T.M. Lenton et al. 2020. Climate change, ecosystems and abrupt change: Science priorities. *Philosophical Transactions of the Royal Society B: Biological Sciences* 375:20190105. https://doi.org/10.1098/rstb.2019.0105.

Udrescu, S.M., and M. Tegmark. 2020. A.I. Feynman: A physics-inspired method for symbolic regression. *Science Advances* 6:eaay2631. https://doi.org/10.1126/sciadv.aay263.

Vandermeer, J.H. 1972. Niche theory. *Annual Review of Ecology, Evolution, and Systematics* 3:107-132. https://doi.org/10.1146/annurev.es.03.110172.000543.

VEMAP. 1995. Vegetation/Ecosystem Modeling and Analysis Project: Comparing biogeography and biogeochemistry models in a continental-scale study of terrestrial ecosystem responses to climate change and CO_2 doubling. *Global Biogeochemical Cycles* 9:407-437. https://doi.org/0.1029/95GB02746.

Volkov, I., J.R. Banavar, F. He, S.P. Hubbell, and A. Maritan. 2005. Density dependence explains tree species abundance and diversity in tropical forests. *Nature* 438:658-661. https://doi.org/10.1038/nature04030.

Wang, H., O.K. Atkin, T.F. Keenan, N.G. Smith, I.J. Wright, K.J. Bloomfield, J. Kattge, P.B. Reich, and I.C. Prentice. 2020. Acclimation of leaf respiration consistent with optimal photosynthetic capacity. *Global Change Biology* 26:2573-2583.

West, G. 2013. Wisdom in numbers. *Scientific American* 308:14. https://doi.org/10.1038/scientific american0513-14.

West, G. 2017. *Scale: The Universal Laws of Growth, Innovation, Sustainability, and the Pace of Life in Organisms, Cities, Economies, and Companies*. New York. Penguin Press.

West, G.B., J.H. Brown, and B.J. Enquist. 1997. A general model for the origin of allometric scaling laws in biology. *Science* 276:122-126. https://doi.org/10.1126/science.276.5309.122.

West, G.B., J.H. Brown, and B.J. Enquist. 1999. A general model for the structure and allometry of plant vascular systems. *Nature* 400:664-667. https://doi.org/10.1038/23251.

Wieder, W.R., C.C. Cleveland, W.K. Smith, and K. Todd-Brown. 2015. Future productivity and carbon storage limited by terrestrial nutrient availability. *Nature Geosciences* 8:441–444. https://doi.org/10.1038/ngeo2413.

Wilkinson, M., M. Dumontier, I. Aalbersberg, G. Appleton, M. Axton, A. Baak, N. Blomberg, et al. 2016. The FAIR guiding principles for scientific data management and stewardship. *Scientific Data* 3:160018. https://doi.org/10.1038/sdata.2016.18.

WMO (World Meteorological Organization). 2021. *State of the Global Climate 2021: WMO Provisional Report*.

Yang, J., B.E. Medlyn, M.G. De Kauwe, and R.A. Duursma. 2018. Applying the concept of ecohydrological equilibrium to predict steady state leaf area index. *Journal of Advances in Modeling Earth Systems* 10:1740-1758. https://doi.org/10.1029/2017ms001169.

4

Research Infrastructure that Enables Continental-Scale Biology

INTRODUCTION

Chapter 3 describes theoretical frameworks that could be applied to addressing questions raised by continental-scale biology (CSB). This chapter identifies the research infrastructure—including tools, networks, and synthesis—necessary to understand the core biological processes related to continental biosphere activities across scales (Figures 4-1a, b). Core biological processes encompass the mechanisms and the nature of change in living systems, including those essential for the growth, development, maintenance, and survival of living systems, across biological scales from molecular and cellular levels to organisms and from organisms to populations to communities to ecosystem levels. These processes affect the structure and functioning of biological systems. To advance the vision of CSB, the research community will need to couple satellite and airborne remote sensing with regional and local observations from the microbial to the macrobial (e.g., animals and plants) worlds to capture processes on a continental scale through networks of expertise, long-term ecological research sites, complex data, and citizen/community science. Through synthesis from a systems perspective, continental-scale observations, models, and experiments can be leveraged to develop process-based understanding that spans organizational, spatial, and temporal scales.

TOOLS

A range of tools can be used to capture information across broad spatial and temporal extents, extend observations from local to continental scales, and connect processes or observations across sites or across scales. Tools include the full range of technologies for observing, quantifying, and interpreting biological patterns and for identifying physical and chemical processes and their underlying mechanisms. Here we emphasize tool-based science encompassing: (a) observational studies that use ground-based

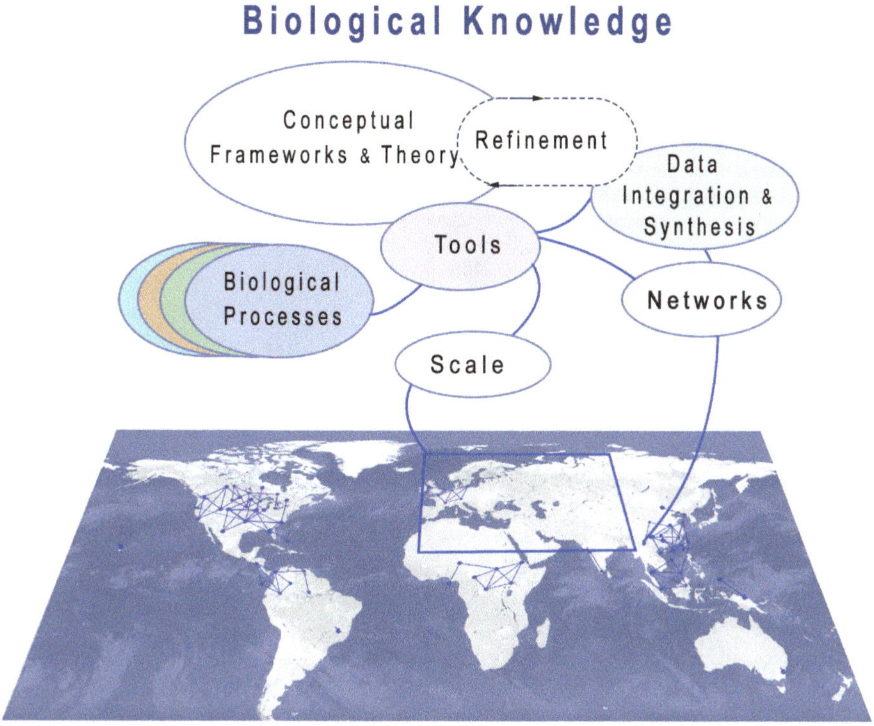

FIGURE 4-1a Relationships among tools, networks, and synthesis centers to biological processes in the context of biological knowledge.
SOURCE: Stacy Jannis.

technology to capture population-level or community dynamics; (b) remote sensing technologies that capture processes at large spatial scales; (c) long-term, ground-based experiments that manipulate variables and are replicated across space; (d) modeling approaches that connect disparate information to enable inferences from observations or predict outcomes over large spatial extents and through time; and (e) tools that support modeling efforts, such as data harmonization, machine learning (ML), and artificial intelligence (AI) approaches.

Observational Studies Using Ground-Based Technologies

Observational studies that employ measurement techniques and allow organizational, spatial, and temporal scales to be traversed or connected are critical for CSB. Some of the most valuable, economical, efficient, and widely used tools in ground-based studies are briefly described below.

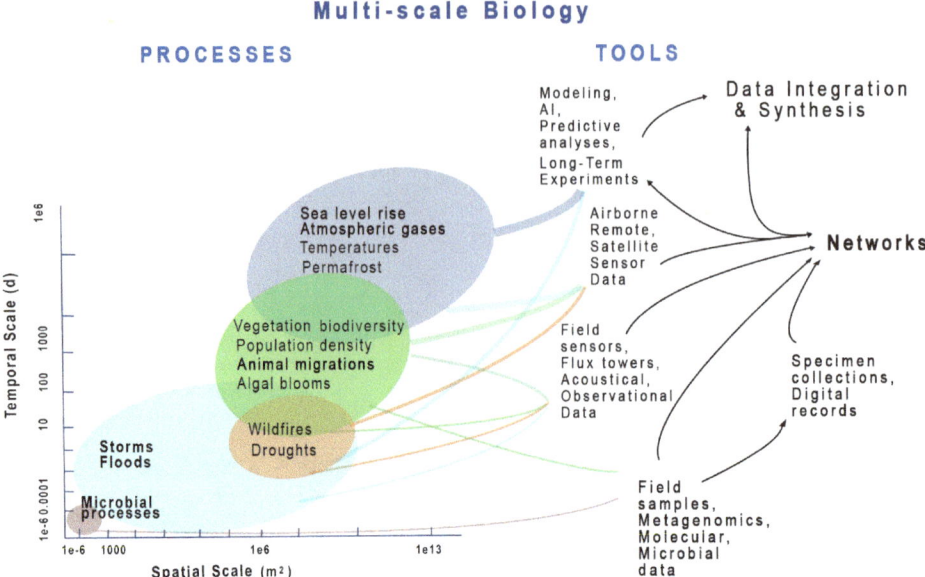

FIGURE 4-1b Temporal and spatial scales of biological and physical processes and patterns in the context of multiscale biology. This includes their relationships to analytical and sampling tools, networks, and synthesis centers as well.
SOURCE: Stacy Jannis.

eDNA and -'Omics

Environmental DNA (eDNA) and multi-omic approaches measure genetic material (e.g., DNA, RNA), proteins, and metabolites in different environmental matrices—for example, sediments, soils, water, air, and plant and animal tissues—to provide insight into microbial and macrobial life dynamics. The growing database of such measurements provides valuable information on the identity, diversity, and function of organisms across scales. Applications include characterizing species distributions and dynamics, the functional dynamics of species, and species interactions and impacts on ecosystem processes.

The distribution and dynamics of macrobes (animals, plants, etc.) can be analyzed through eDNA and eRNA techniques. eDNA proves essential in monitoring animal populations, providing insights into their diet, health, and evolutionary paths that influence their survival and resilience in a changing climate (Grieneisen et al. 2021). These approaches generally use targeted analysis of a single gene and then phylogenetic analysis to determine which macrobial taxa these sequences belong to. These gene targets can be sampled at point locations, but as material tends to distribute or be left behind by moving animals, they represent very large spatial areas, extending observation capabilities into previously inaccessible or difficult-to-access ecosystems, including wetlands, freshwater, coastal, and marine systems. eRNA offers the possibility of analyzing the expressed functions and levels of stress of organisms, and/or even the genomics of RNA viruses.

For microorganisms, molecular approaches used to understand the roles, interactions, and functional dynamics across extensive spatial and temporal scales have been fundamental to the application of so-called multi-omics approaches. One of these approaches, "amplicon sequencing," has been used by the Earth Microbiome Project on over 200,000 samples to catalog continental-scale microbial diversity (Thompson et al. 2017). However, amplicon sequencing, as with most eDNA approaches for macrobes, provides information only on the taxonomic distribution of microorganisms.

For analysis of the distribution of microbial genomes and their functional traits we deploy metagenomics, metatranscriptomics, MetaRiboSeq, metabolomics, and metaproteomics (see Box 4.1). Those techniques provide vast data resources that can facilitate robust hypothesis testing to explore the ecology of microorganisms and their impact on hydrological dynamics, nutrient cycling, and climate active atmospheric processes. For example, 'omics can be used to study the role of soil-associated fungi and bacteria, such as Verrucomicrobia, in the uptake of carbon in grasslands (Brewer et al. 2017, Fierer et al. 2013), which, in turn, can provide insight into how plant distributions at local and regional scales influence soil microbial carbon dynamics. For aquatic ecosystems, multi-omic techniques elucidate microbial community dynamics in response to significant events such as oceanic current shifts or river diversions, supported by databases such as the Genome Resolved Open Watersheds database (Borton et al. 2023).

BOX 4-1
Connecting the Tools and Networks That Enable CSB to Its Core Themes

In examining the tools and networks that enable the study of CSB, it is vital to reconnect to the established core themes of this field, as detailed in Chapter 2: Biodiversity and Ecosystem Function, Resilience and Vulnerability, Connectivity, and Sustainability of Ecosystem Services. Tools and networks are at play within each of these themes. For example, the National Ecological Observatory Network (NEON, Figure 4-2) couples remote sensing and standardized ground-based measurements across sites to help understand how biodiversity and ecosystem function are related, while researchers use tools such as modeling to evaluate current and future sustainability of ecosystem services. Observational studies across time and space, including methods such as measuring atmospheric gases, contribute toward measuring factors such as land-use change and human activities that relate to resilience and vulnerability. Similarly, the array of tools, networks, and synthesis centers that this chapter describes provide critical support for understanding connectivity among ecosystems and among coupled human and natural systems. This non-exhaustive list exemplifies how intrinsic different tools and networks contribute toward the underlying themes of CSB.

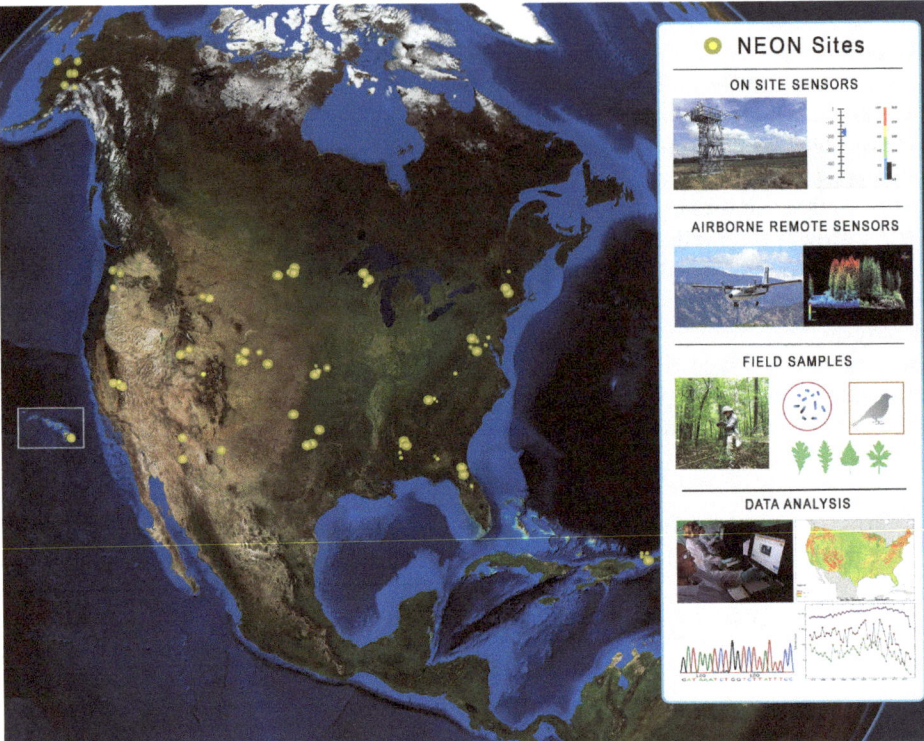

FIGURE 4-2 National Ecological Observatory Network. NEON sites are represented by yellow dots across North America. These sites collect data via numerous methods, listed and represented via images on the right. These methods include on-site sensors, airborne remote sensors, field sample collection, and a variety of methods for data analysis. The structure and coordinated methods for collection allow these data to be useful across multiple temporal and spatial scales.

Plant genomics, RNAseq, and metabolomics are powerful tools that collectively enable a comprehensive understanding of plant processes and dynamics on a continental scale. Plant genomics provides insights into the genetic blueprint of various plant species, revealing the diversity and evolutionary adaptations across different ecosystems. RNAseq offers a detailed view of gene expression patterns, allowing researchers to identify how plants respond to environmental stressors, pathogens, and climatic changes at a molecular level, and how plant species and disturbances such as wildfire affect the microbial composition of soils (Osburn et al. 2021, VanderRoest et al. 2024). Metabolomics complements these approaches by profiling the biochemical compounds produced by plants, shedding light on metabolic pathways and the functional outcomes of genetic and transcriptomic variations. Together, these techniques facilitate the elucidation of complex plant–environment interactions, the discovery of novel genes and pathways involved in adaptation and resilience, and the development of strategies for enhancing crop productivity and sustainability across diverse biomes. This integrative approach is crucial for addressing global challenges such as food security, climate change, and ecosystem conservation.

Some unique challenges exist with eDNA and eRNA techniques. They need to be calibrated based on body mass or genomic polyploidy to ensure accurate quantitation of the target organism. They also need to take into consideration the degradation rates of DNA and RNA in different environmental matrices and contexts. Finally, they are affected by sampling bias, with fluid matrices (e.g., streams and rivers) having a greater distribution potential for biomarkers, and spatial analysis. Most multi-omic data are currently proportional, which limits the opportunities for continental-scale integration of multiple sites or temporal scales. Because of these limitations, multi-omic data are not sufficiently integrated into models. Recent advances in quantification need to be further expanded to create quantitative multi-omic resources and fully integrate these data into models that inform policy.

Isotopes: Stable and Radioactive

Stable and radioactive isotopes are one of the most powerful tools we have to resolve biological processes across scales. For example, radioactive carbon isotopes (^{14}C) are being used to resolve the mean age and transit times of carbon in terrestrial systems. In the mid-to-late-1950s, the prevalence of atomic bomb tests and use increased the amount of radioactive ^{14}C in the atmosphere. This pulse of radiocarbon, along with known decay rates and discrimination processes, allowed organic material to be dated and tracked through ecosystems with relatively high precision (Hasler 2022). This tool has helped us understand that the mean age of carbon is much greater than the mean transit time of carbon in terrestrial systems, and that the mean age of soil carbon in tropical systems is an order of magnitude younger than the soil carbon in permafrost regions, both of which constrain and inform soil carbon loss and sequestration in Earth system models (Shi et al. 2020).

Combining classical metabolomics, which focuses on metabolite levels, with substrates that are labeled with stable isotopes can yield insights into metabolic flux and help resolve metabolic rate and flow (Yu et al. 2023). For example, ammonia and nitrate labeled with ^{15}N coupled with metabolomics can quantify the rates of and pathways used by plants to assimilate the nitrogen from soil that they need for growth (Kurczy et al. 2016). As another example, stable carbon isotope labeling in trees experiencing prolonged drought can demonstrate that belowground tissues store and use carbohydrates preferentially over aboveground tissues during recovery from drought (Hagedorn et al. 2016).

Some challenges exist with isotopic approaches. Radioactive carbon isotopes (^{14}C) face challenges of decay and dilution. Fossil fuel emissions, which produce a large amount of CO_2 with no ^{14}C signal because fossil fuels have lost all ^{14}C over millions of years of radioactive decay, are thus diluting the ^{14}C tracer. Atmospheric CO_2, and therefore newly produced organic material, will appear as though it has "aged," or lost ^{14}C by decay. By 2050, fresh organic material could have the same $^{14}C/^{12}C$ ratio as

> **BOX 4-2**
> **Application of Omics Technologies**
> **That Can Be Applied to CSB**
>
> - *Metagenomics* deciphers collective genetic material from environmental samples, offering a panorama of the microbial species (including bacteria, archaea, eukarya, and viruses) within an environment as well as cataloging their potential functions (Nayfach et al. 2021). For instance, analyzing soil samples from various environments reveals microbes adapted to those conditions, potentially uncovering evolutionary drivers such as those influencing carbon dynamics in watersheds (Long et al. 2016).
> - *Metatranscriptomics* focuses on RNA, specifically messenger RNA, to determine active gene expression within a sample or to analyze RNA viruses. This approach reveals real-time microbial responses to environmental changes, such as the expression of genes involved in nutrient influx or changes in diurnal gene expression related to photosynthetic cycles.
> - *MetaRiboSeq* is a relatively new technology that actively isolates the messenger RNA associated with ribosomes in cells. These techniques allow us to capture and sequence the messenger RNA as it is in the process of being translated into amino acids to make a protein. The messenger RNA pool represents transcriptomic potential, while the riboseq mRNA data represent active translation of that pool (Fremin et al. 2020).
> - *Metaproteomics*, utilizing high-resolution mass spectrometry (MS), identifies and quantifies proteins in microbial communities, providing insights into active metabolic processes and pathways. For example, it can pinpoint the enzymatic machinery involved in pollutant degradation, aiding in precise bioremediation strategies (Püttker et al. 2015).
> - *Metabolomics* characterizes small molecules, either within a cell or present outside of the cell, offering insights into microscale metabolic reactions occurring across ecosystems. This analysis, using MS and nuclear magnetic resonance (NMR) spectroscopy, helps to characterize how microbial communities respond to larger environmental drivers such as climate change and land-use alterations.

samples from 1050, and thus be indistinguishable by radiocarbon dating. As Graven (2015) notes, "Some current applications for ^{14}C may cease to be viable, and other applications will be strongly affected." Dating using ^{14}C from solar proton events may offer a solution (e.g., Walker et al. 2023).

Atmospheric Gases, Flux Towers

Measures of atmospheric greenhouse gases—such as carbon dioxide, methane, and nitrous oxides—and water vapor, provide critical insights into ecosystem dynamics, carbon cycling, and climate feedback mechanisms. They enable monitoring of the influence of biological variation, human activities, land-use change, and climate variations on the biosphere dynamics. Atmospheric gas concentrations have a long history of measurement, from flask samples to flux towers, where analyzers are used in combination with three-dimensional (3D) sonic anemometers to estimate gas flux. Flux towers facilitate the measurement and detection of gas exchanges between the biosphere and the atmosphere, serving as an essential tool for assessing how biological activities influence and are influenced by atmospheric changes. For example, towers measuring carbon and water fluxes over Northern Hemisphere forests demonstrated that these forests have increased their carbon gain per unit of water used over a 20-year period (Keenan et al. 2013). Systematic analyses of various competing hypotheses to explain this trend indicated that the observed increase is most consistent with a strong CO_2 "fertilization" effect, whereby trees are growing more due to more carbon substrate in the air.

Flux towers are important for scaling between terrestrial and atmospheric processes. For example, to gain insight into Earth's metabolism and how it is changing in response to increasing surface temperature, networks of long-term measurements of atmospheric CO_2 concentrations have been combined with Earth system models to resolve why the Northern Hemisphere's atmospheric (CO_2) seasonal amplitude (the difference between summer and winter CO_2 levels) has increased with surface warming. Forkel et al. (2016) found climate-warming stimulated plant carbon uptake faster than respiratory carbon release from the terrestrial biosphere, although seasonal respiration processes are influenced by a range of interacting drivers including snow, permafrost, vegetation composition and structure, drought, soil properties, and fire disturbance history (Chylek et al. 2022, Previdi et al. 2021, Rantanen et al. 2022, Treat et al. 2024). Efforts to further incorporate these process-level interactions are needed to better resolve both spatial and temporal variations in atmospheric CO_2 integrated across vast spatial scales. Other examples using flux towers combined with peripheral sensors, such as those that measure volatile organic compounds, are measuring forest stress from disturbances such as drought and insect outbreaks (Kravitz et al. 2016).

In specific cases where a flux tower or flask sampling is absent, biological samples can serve as indices for air quality. For nearly 25 years, the U.S Department of Agriculture (USDA) Forest Service's Forest Inventory and Analysis (FIA) program collected lichens and recorded lichen richness on a subset of standardized plots throughout the United States (Jovan et al. 2021). These vouchered samples and data can serve as a proxy for air quality when they are validated or related to quantitative measures of air quality. For example, the relative dominance of lichen functional groups can be directly related to nitrogen deposition and be used to estimate an ecosystem's critical load (Root et al. 2015).

Physical and Digital Samples and Collections That Inform Biodiversity

Biodiversity collections housed in natural history museums and herbaria provide the foundations of our knowledge and documentation of life on Earth (NASEM 2020). Preserved specimens capture species variation over time and space and are important to maintain because they allow us to study functional traits of organisms in common environments (Fontes et al. 2022, Perez et al. 2019). Increasingly, botanical gardens, zoos, aquaria, seed collections, and insectaria are hubs of ex situ conservation efforts (Westwood et al. 2021, Wood et al. 2020). A wide range of efforts has been undertaken to enhance such collections, including digitization, aggregation of digital records into databases, enhancement of collections by citizen science, and collections of nonphysical specimens such as images or recordings.

Digitization of physical collections is an ongoing and important effort (NASEM 2020). A major effort by the National Science Foundation (NSF) to enable the digitization of specimens across the tree of life has been ongoing for the past decade.[1] These efforts are directly in line with NSF's current emphasis on Innovative Use of Scientific Collections, which includes the Directorate of Biological Sciences (Division of Integrative Organismal Systems, Division of Biological Infrastructure, and Division of Environmental Biology), aimed at fostering innovative and diverse uses of collections and/or associated digital data for novel research and training applications. These data have provided the foundation for extended specimens and next-generation digitization efforts (i.e., digitizing additional data from original specimens). Extended specimens provide additional information derived from physical specimens that can be used to understand functional and ecological attributes of organisms. These include foliar spectral data from physical leaf samples (Kothari et al. 2023, Meireles et al. 2020), which can provide estimates of forest ecosystem function, and CT (computed tomography) scans of skeletons (Poo et al. 2022, Shi et al. 2018). These next-generation spectroscopic data (reflected or transmitted light across many wavelengths, typically 400–2,500 nm) collected from foliar samples can enable prediction of chemical and other functional traits of plants, such as leaf nutrient content predicting photosynthesis (Kothari et al. 2023, Serbin et al. 2014, Singh et al. 2015).

Digital records can then be aggregated into databases for multiple uses to connect across scales. Specimen records are incorporated by aggregator nodes across the globe that input species occurrence records and geographic coordinates and uncertainties into the Global Biodiversity Information Facility (GBIF[2]). In effect, these collections are a proxy for species distributions or trait information, provided the label information on the specimens and geographic data are accurate (Zizka et al. 2020). For species, researchers can then validate these proxies of species distributions with species distribution models, and merge with remotely sensed species occurrences, to improve estimates of changing population size and abundance in the face of global change (Cavender-Bares et al. 2020, Fretwell and Trathan 2021, Guzmán et al. 2023). For traits, such as plant functional trait data in databases such as the Plant Trait Database (TRY) and the Botanical Information and Ecology Network (BIEN) (Enquist et al. 2009, Kattge et al. 2020), researchers can link these to spectroscopic remote sensing (see below) to map functional trait variation in vegetation to provide critical habitat and ecosystem information. Similarly, in soil systems, microbial functional trait analysis is gaining prominence. Akin to the aggregation of plant trait data in TRY and BIEN, microbial trait data are pivotal in understanding soil microbial communities and their impact on ecosystem health (Barberán et al. 2015,

---

[1] See https://www.nsf.gov/awardsearch/showAward?AWD_ID=2027654 (accessed May 8, 2024).
[2] See gbif.org.

Buzzard et al. 2019, Wieder et al. 2015, Yang 2021). The integration of these microbial data with advanced spectroscopic techniques, which predict chemical and functional traits of plants, offers a comprehensive view of ecosystem functions and can be linked to remote sensing for habitat and ecosystem mapping.

Physical collections can be extended by citizen/community science efforts. Efforts such as iNaturalist[3] enhance biological collections by providing a wealth of additional species occurrence records, greatly expanding the spatial extent and amount of available data. Despite the use of online training on data collection and protocols in community science efforts, there can be uncertainties and other problems with the collected data. Other collection approaches are coupled with digital libraries to train AI models that classify the identity of organisms with increasing accuracy (Vélez et al. 2023). These include camera trap images and acoustic recordings (e.g., Clark et al. 2023, Quinn et al. 2023).

Acoustic recording units (ARUs) facilitate the passive observation of birds and wildlife. Acoustic technologies are rapidly developing to obtain species occurrence and animal migration patterns at scales where monitoring has not previously been possible. Acoustic libraries of birds, for example, are well developed and are used to train models for identification with high accuracy (Kahl et al. 2021, Ruff et al. 2023). Similarly, camera traps provide a means to capture local occurrence and behavior records that enable population information on animals. These cameras are equipped with a motion sensor, usually a passive infrared sensor or an active infrared sensor using an infrared light beam and are automatically triggered by a change in activity in the vicinity, such as animal motion (O'Connell et al. 2011). As with ARUs, images of species obtained from camera traps can be automatically identified using ML models, facilitating efficient analysis pipelines (Tabak et al. 2019). One approach coupled with Zooniverse[4] crowdsources the identification of images, compared to expert identification, to develop AI identification from camera images (Willi et al. 2019).

Finally, tracking of animal movement and migrations have advanced the study of a wide range of wildlife, including birds, zebras, elephants, caribou, and many other species. Collection of these data into large, organized databases such as MoveBank (Kays et al. 2022) have advanced this research substantially (Davidson et al. 2020, Tucker et al. 2018). The International Cooperation for Animal Research Using Space (ICARUS) antenna on the International Space Station received signals from tiny transmitters attached to over 800 species of animals ranging from bats to elephants to track migration patterns, but unfortunately in March 2022 was unexpectedly terminated, impacting our ability to assess the influence of biodiversity on ecosystem function. This functionality is expected to resume with the resurrection of ICARUS on small satellites (CubeSats) in 2025. Other future advances in animal tracking and expanded use of drones have potential to augment the resumption of tracking data from ICARUS.

---

[3] See https://www.inaturalist.org/.
[4] See https://www.zooniverse.org/.

Satellite Observations and Remote Sensing Campaigns Coordinated Across Space and Time and Integrated with Ground-Based Measurements

Earth observing satellites and airborne remote sensing by aircraft and drones greatly facilitate the ability to scale across both time and space. Observations can span daily, weekly, monthly, seasonal, annual, and decadal temporal scales and be captured at sub-meter spatial scales (e.g., airborne sensors such as those used by NEON) to regional, continental, and even global extents (satellites). There are many combinations of these temporal and spatial scales depending on the data source, but in general, finer spatial resolution from satellites requires a longer cadence (i.e., the time between repeat coverage of any given area), except in the case of some microsatellite constellations or sensors that can be tasked to frequently target specific locales.

For terrestrial ecological observations, the measurements acquired by both satellites and aircraft need to be transformed into data that are meaningful from an ecological perspective. For example, optical reflectance measurements across a range of wavelengths (i.e., visible to infrared), such as those from the moderately high spatial resolution (30-m) Landsat series of satellites, have been used for decades to estimate the photosynthetically active radiation absorbed by plant canopies (Zeng et al. 2020) and to classify vegetation into land cover and/or plant functional types based on their composition and phenology (Hansen et al. 2013, Potapov et al. 2022, Zeng et al. 2022). Other moderate-resolution optical satellites, such as the Moderate Resolution Imaging Spectroradiometer (MODIS) sensors on the Terra and Aqua satellites, have been used to map vegetation composition, phenology, and type classifications, and also to quantify gross and net primary productivity and other functional attributes of terrestrial vegetation (Ryu et al. 2019, Xiao et al. 2019).

Additional efforts that have enhanced the use of remote sensing information to inform CSB include hyperspectral imagery and light detection and ranging (lidar). Remote sensing instruments that capture reflectance across many wavelengths with very high spectral resolution (i.e., hyperspectral) have been used to model plant functional traits based on foliar trait modeling methods (e.g., Wang et al. 2023) and to map community composition and plant disease or stress (e.g., Guzmán et al. 2023). Remotely sensed measures of functional, structural diversity, and composition maps can be used to determine changing aspects of ecosystems including diversity and function (Wang and Gamon 2019, Gholizadeh et al. 2020, Laliberté et al. 2020, Liu et al. 2024) and even to predict belowground ecosystem processes (Cavender-Bares et al. 2022, Lang et al. 2023). These can also be coupled with other measures of composition and diversity, including eDNA, other 'omics approaches, and flux towers (Box 4-3).

Lidar instruments are useful for mapping of 3D structural attributes of vegetation, including foliage height profiles, plant area volume density, canopy height, and aboveground biomass, among others. A novel instrument on the International Space Station, the Global Ecosystem Dynamics Investigation (GEDI) has been used to map, for the first time, forest 3D structural properties with robust error and uncertainty estimates (Dubayah et al. 2020, 2021). Before GEDI, 3D structural metrics of vegetation were prone to large error and uncertainty, and it was only possible to map limited spatial and

BOX 4-3
Application of Remote Sensing to Collect Vegetation Spectral and Structural Data Relevant to CSB

Airborne spectroscopic imagery and lidar waveforms from the International Space Station and an Earth observing satellite detect aboveground ecological attributes that predict belowground properties and processes. Remotely sensed spectroscopic imagery can be used to predict (1) foliar chemistry and traits (e.g., specific leaf area, leaf C, leaf N, sugars, hemicellulose, cellulose, lignin), (2) functional and phylogenetic composition of vegetation (e.g., legumes, C4 grasses) and (3) aboveground productivity. These vegetation attributes in turn predict belowground properties and processes (e.g., microbial biomass N and C, net N mineralization rates, soil carbon, enzymatic breakdown of litter).

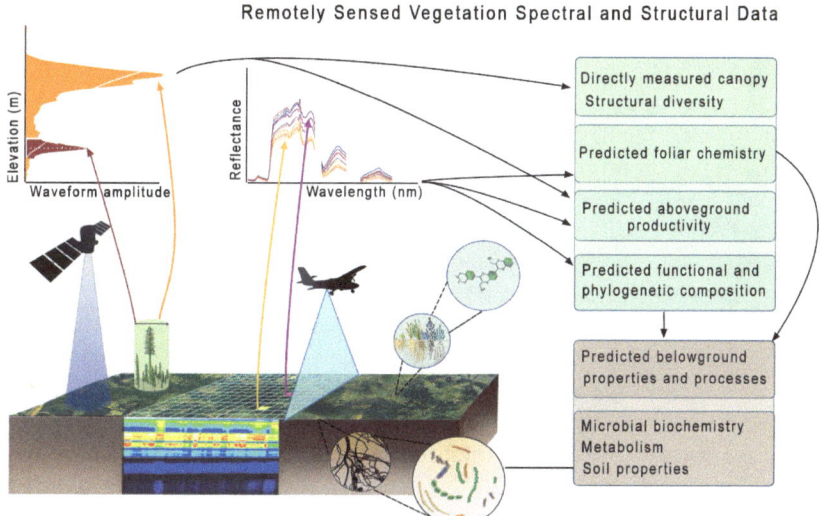

FIGURE 4-3-1 Image of airborne platform using full-range spectroscopy (400–2,500 nm) and remotely sensed lidar (shown here from satellite). Each pixel of the spectroscopic image data cube has a unique spectral reflectance fingerprint containing vast information about the chemical, functional, and structural attributes of vegetation. The lidar-generated waveform directly measures aboveground structure and structural diversity and predicts productivity. Machine learning and statistical models can be developed from the spectroscopic imagery to predict aboveground functional traits of vegetation (Asner and Martin 2008, Miraglio et al. 2023, Wang et al. 2020), plant diversity (Gholizadeh et al. 2019, Laliberté et al. 2020, Pinto-Ledezma and Cavender-Bares 2021, Wang et al. 2018), phylogenetic composition (Griffith et al. 2023) and forest productivity (Williams et al. 2020), as well as belowground properties and microbial processes (Cavender-Bares et al. 2022, Sousa et al. 2021). Predicting belowground properties from remote sensing imagery is possible due to mechanistic linkages between aboveground and belowground portions of ecosystems.
SOURCE: Stacy Jannis.

temporal extents using airborne lidar data acquisitions, such as those systematically collected over NEON sites (Hakkenberg et al. 2023). These recent lidar-based 3D structure metrics and associated data products also provide valuable calibration information, particularly when they themselves are calibrated with field data (which is the case with GEDI lidar products), to map and model 3D structural property metrics over continental to global spatial extents (Duncanson et al. 2022, Ma et al. 2023).

Large-scale, coordinated remote sensing campaigns from government agencies, such as NASA, have been central to the vision for CSB, that is, conducting coordinated research across scales to provide novel insights that would not otherwise be possible. NASA campaigns and missions have evolved from the First International Satellite Land Surface Climatology Project (ISLSCP) Field Experiment, which focused on the Konza Prairie in Kansas during the 1980s (Sellers et al. 1988); to the Boreal Ecosystem-Atmosphere Study (BOREAS) which targeted a gradient of boreal forest sites in central Canada during the 1990s (Sellers et al. 1995, 1997); the Large-Scale Biosphere-Atmosphere Experiment in Amazonia (LBA) which focused on Amazonian tropical forests during the 1990s and 2000s (Avissar and Nobre 2002, Avissar et al. 2002); and most recently the ongoing Arctic-Boreal Vulnerability Experiment (ABoVE) which is targeted on the tundra and taiga forest biomes of Alaska and western Canada; and the Department of Energy's Next-Generation Ecosystem Experiments in the Arctic (NGEE Arctic) and the Tropics (NGEE Tropics). The BioSCAPE mission in South Africa and others currently being scoped will span tropical wet and dry forest regions in Central and South America, including in the Amazon and the Brazilian Cerrado, and West Africa (PAN tropical investigation of bioGeochemistry and Ecological Adaptation [PANGEA]), and separately across a range of arid land ecosystems (Actionable Science for Earth's Changing Drylands [ARID]).

These coordinated campaigns, augmented with more localized manipulation experiments such as the Spruce and Peatland Responses Under Changing Environments (SPRUCE), provide valuable insights and serve as models for continental-scale research. The campaigns have allowed some of the best examples of CSB, including documenting how disturbances interact to produce a range of ecosystem states that persist for decades (Foster et al. 2022), how fire disturbance influences vegetation succession and associated feedbacks to climate through changing surface reflectivity (Massey et al. 2023), how the carbon budget of high latitudes has changed in recent decades (Wang et al. 2021, Watts et al. 2023), where methane sources arise across wetland landscapes and as a result of fire (Yoseph et al. 2023), how trees have been expanding their northern range limits in recent decades (Dial et al. 2022), where the boreal forest is becoming more and less productive (Berner and Goetz 2022), and how wildlife, including beavers and moose, are expanding their ranges and altering Arctic ecosystems (Tape et al. 2022), among many other examples.

Remote sensing campaigns and Earth observation data face challenges associated with continuity and storage. As technology advances and campaigns are discontinued and others launched (e.g., ICARUS), retrospective research to ensure continuity of observations through time or that allows interoperability through statistically robust documentation of directional change (e.g., via time series analysis) will be challenging. This may require aggregating the spatial and temporal resolution of recent data to match that of data collected in prior years and decades. The sheer volume of data may also be a challenge for many users of these data due to computing demands (storage and processing speed). NSF's CyVerse at the University of Arizona (formerly the iPlant Collaborative) was designed to create the capacity to address some of these challenges with its high-performance computing capabilities and cloud storage of large datasets.

Long-Term Experiments—Longitudinal in Time, Replicated in Space

Manipulative experiments are those that control specific variables that allow us to determine the causal factors that drive observed processes. Relevant to enabling CSB are experiments that control key variables, such as plant diversity, climate, and greenhouse gases. Experiments that control these variables have been critical in deciphering the impacts of global change factors on organism function, community assembly processes, and ecosystem function (Blondeel et al. 2024, Isbell et al. 2015, Kolton et al. 2022, Pastore et al. 2021, Pellegrini et al. 2021, Reich et al. 2018). Distributed long-term manipulative experiments that are replicated across ecosystems and biomes enable us to examine how the same causal drivers of change have consistent and heterogeneous impacts in space and time.

These distributed manipulative experiments often become organized networks (see section below), for example, Nutrient Network (NutNet) (Box 4-4), Drought Network (DroughtNet), and Tree Diversity Network (TreeDivNet) (Table 4-1). Control replicates (i.e., replicates that do not have treatments imposed) from long-term manipulative experiments by themselves often become some of the most valuable observational studies—which are located at specific sites and replicated in space and/or time. The results of these experiments can reveal global change feedbacks in changing environments on ecosystems and change our understanding of synergistic impacts of different factors (e.g., the Biodiversity, CO_2, and Nitrogen [BioCON] experiment (Reich 2009).

As with the other tools described above, long-term experiments experience myriad challenges. Investigators who measure any one variable for decades, such as air humidity, can experience changes in both theory and equipment based on that theory (Sonntag et al. 2021). For example, early measurements of air humidity at Coweeta Hydrologic Lab, a USDA Forest Service Experimental Forest that was established in the early 1930s, used hair hygrometers, while more recent measurements use electrodes and optical measurements (Miniat et al. 2021). When sensors and theory do change during the course of a long-term experiment, conducting simultaneous measurements with both sets of sensors over the full range of conditions is always desirable, but rarely available or affordable. Challenges with continuity of measurements, including human and financial resources, are thus almost always a challenge in long-term experiments.

Models to Infer Process and Pattern: Physical, Hierarchical, Empirical, Statistical, Process-Based, Earth System Models, and Species Distribution Models

Modeling is needed for CSB, because ecological systems are often too large and slow moving for hypotheses to be tested at the relevant temporal and spatial scales through individual experiments. Ecological principles underscore CSB, recognizing the inherent hierarchy within ecological data. Hierarchical modeling provides multiscale insights, integrating data on soil microbial communities, soil chemistry, and climate (Fierer et al. 2012). In a hierarchical model, large, complex stochastic systems can be represented by a sequence of smaller probabilistic parts. As a simple example, many questions ecologists ask are about population size. As population size is affected over time by factors such as survival and recruitment, it is useful to look at population size one hierarchical level up to look at multiple populations in different locations, often termed "metapopulations." Numerous species distribution models have been developed to predict the distribution of populations and species across space (Franklin 2010, Frans and Liu 2024).

A process-based model represents one or more processes in a well-defined biological system, for example, models of biochemical pathways or population dynamics models. Process-based modeling, including Earth system models (ESMs), amalgamates diverse information, crucial for forecasting ecosystem reactions and understanding functional ecosystem dynamics (Prosser 2015). Addressing variability in microbial dynamics, spatiotemporal sampling, along with interpolation and extrapolation techniques, predicts microbial assemblies and spatial structures. One example of this prediction is seen in an investigation of coastal microbial dynamics in the English Channel, where a dense longitudinal time series from one location was leveraged to create a neural network that predicted microbial community composition from another site. This neural network was validated on data not used to train it, and then used to interpolate and extrapolate predictions on community structure across the whole English Channel over a period of 10 years (Gibbons et al. 2013, Gilbert et al. 2012, Larsen et al. 2015). The same ecological extrapolations have been deployed for soil systems to capture and predict shifting continental-scale processes (Fierer et al. 2013, Ladau et al. 2018).

The synthesis of 'omic data allows extrapolation from localized niches to broader expanses, presenting a clear view of overarching biological patterns, including changes over time. Even though the application of multi-omics in predicting emergent properties across ecosystems is in its infancy, it holds promise for understanding organism responses to environmental factors at various scales. Multi-omics can highlight plant–microbe interactions, plant adaptations to stress, or microbial biogeochemical processes across different terrains and climates. For instance, it can predict microbial activity's role in forests' response to drought or wetlands' changes with increased salinity. In ESMs, multi-omics help in understanding the microbial pathways in carbon sequestration across continents and refining global carbon cycle models. However, understanding the feedback mechanisms in these processes is essential for future extrapolations.

Quantifying these mechanisms is still a challenge, with limited evidence outside of medicine for the effective use of multi-omics in this context.

Tools to Support Modeling for CSB: ML, AI, and Data Harmonization Centers

Approaches to facilitate data integration and interpretation include the application of AI techniques, for example, ML analysis to identify traits that predict variance in key parameters. These approaches can be used to identify features that can be integrated into hierarchical modeling to capture the dynamic processes that underpin emergent properties of the complex systems that make up ecosystems. Leveraging burgeoning multi-omic datasets necessitates AI and ML techniques to identify dynamics and ecosystem disturbances. Challenges in the application of AI and ML are described further in Chapter 3.

Distributed Active Archive Centers (DAACs) work to harmonize data. The Oak Ridge National Laboratory's DAAC publishes and preserves NASA data relevant to terrestrial ecology, primarily field and airborne data. They also facilitate the use of NASA data in ways that address terrestrial ecology needs. They host over 1,700 datasets across 9 science themes and 36 missions and projects. They make large and complex remote sensing data easily accessible to researchers focusing on smaller scales of organization (e.g., organisms, plots), and allow integration of diverse and discrete ecological data at smaller scales and connecting them to large remote sensing data. One example is the soil moisture visualizer, where soil moisture datasets collected across observational networks (e.g., USDA Soil Climate Analysis Network (USDA SCAN), SNOw TELmetry Network (USDA SNOTEL), and NASA's Soil Moisture Active Passive datasets) are aggregated and harmonized and made available to users (Shrestha and Boyer 2019). Data aggregation, harmonization, and visualization allow users, in this case, to make inferences on soil moisture availability or drought across wide expanses that eclipse the spatial extent of any one network.

Digital twin platforms, emerging technologies that integrate computer science, mathematics, engineering, and life sciences, offer transformative potential for CSB research (de Koning et al. 2023, NASEM 2024). These platforms can create high-fidelity, real-time digital replicas of biological systems, enabling researchers to conduct virtual experiments and simulations that are impractical or impossible in the real world. This capability is particularly valuable for studying large-scale biological processes and ecosystems. For example, digital twins can model complex interactions within ecosystems, providing insights into the impact of environmental changes, such as climate change or habitat destruction, on biodiversity. Researchers can use these models to test hypotheses, predict future scenarios, and develop conservation strategies. Moreover, digital twins can facilitate data integration from various sources, including satellite imagery, sensor networks, and field observations, helping researchers identify patterns and trends that might not be apparent from smaller-scale studies.

By leveraging advanced computational techniques such as ML and AI, digital twin platforms can analyze vast amounts of data, identify relationships, and predict outcomes with high accuracy. This approach, linked with theory (Chapter 3, Figure 3-1-1), will enable the identification of gaps in knowledge and data collection, enabling hypothesis testing and refinement of models and theory that will lead to robust predictive power for ecological outcomes associated with climate change or ecosystem disturbances. This capacity is crucial for understanding the complexities of continental-scale biological phenomena, where traditional analytical methods may fall short.

NETWORKS

Scientific observational and experimental networks are an organized group of people, knowledge systems, infrastructure, and data focused on central scientific questions or goals that conduct experiments and/or observations across space and/or time (see Table 4-1 for some examples). Networks may standardize measurements across locations, or if they are not standardized but measure the same process, they can at least provide a way to scale or link different measurements or tools together to allow broad-scale inference. Successful networks establish transparent and inclusive policies for participation as a member, collaboration and publication, and the use and reuse of data (NRC 2015, Wilkinson et al. 2016). To promote collaboration and scientific advancements, networks often adopt open science and team science principles (NRC 2015). Networks involve coordination and communication across people, and although data are one of their outputs, they are not the only product (e.g., events, data visualizations, data management and oversight, AI/ML experts, natural history experts).

Many networks are established to answer a basic question or fulfill a mandate, for example: What is the U.S. forest inventory (USDA Forest Service FIA)? How and where is acid deposition affecting air and water quality (National Atmospheric Deposition Program [NADP] and the Clean Air Status and Trends NETwork [CASTNET])? What are the material biospheric fluxes of energy, carbon, and water across the U.S. terrestrial ecosystems' boundary layer (U.S. Department of Energy's [DOE's] Ameriflux network)? How general is our current understanding of productivity–diversity relationships (NutNet)? How are the world's forests changing in biodiversity and carbon (ForestGEO network)? In other words, they are question driven. Some networks, though, are not established to answer a particular question; rather, they function more as data collection or observational networks (e.g., NEON).

Similar to the varied reasons that networks are established, *how* networks are established also varies. Some are established in a top-down fashion; for example, networks are established by government mandate (e.g., FIA and NADP mentioned above), to accomplish a government agency's mission (e.g., USDA Forest Service Experimental Forest and Range network, DOE's Ameriflux, National Oceanic and Atmospheric Administration's U.S. Climate Reference Network), or to coordinate experiments and/or observations (e.g., NEON, Forest GEO, NutNet, DroughtNet, International Tundra EXperiment). Often, smaller networks can grow into larger and more formal networks. The Research Coordination Networks (RCNs) program under NSF has been a very successful tool in establishing networks and expanding smaller networks. For example, NutNet was initiated to address two key human influences on ecosystems—global nutrient augmentation and changes in consumer dynamics (Box 4-4). Networks can

TABLE 4-1 Scientific Observational and Experimental Networks

Network Name	Primary Network Type	Website	Data Description	Temporal Extent	Temporal Resolution	Spatial Extent	Spatial Resolution	Funding Mechanism(s)	Notes
eBird	Citizen/community science	https://ebird.org/home	Location and occurrence data on birds are aggregated to create maps of bird range, abundance, habitat, and trends. Habitats: terrestrial, freshwater, marine.	Trends data 2007–present	Instantaneous	Global	Point location data	Cornell Lab of Ornithology, US National Science Foundation (NSF), Leon Levy Foundation, and others	
iNaturalist	Citizen/community science	https://www.inaturalist.org	Biotic observations connected with geographic location. Habitats: terrestrial, freshwater, marine.	2008–present	Seconds to decade	Global	Depends on spatial accuracy of observation	Nongovernmental organizations (NGOs)	Research-grade observations incorporated into Global Biodiversity Information Facility (GBIF)
North American Breeding Bird Survey (BBS)	Citizen/community science	https://www.pwrc.usgs.gov/bbs/	Biotic (bird observations and counts along permanent BBS routes); abiotic descriptors (weather on observation day). Habitats: terrestrial, freshwater.	1966–present	Annual	Continental	400-m radius around point observation	U.S. Geological Survey	

continued

TABLE 4-1 Continued

Network Name	Primary Network Type	Website	Data Description	Temporal Extent	Temporal Resolution	Spatial Extent	Spatial Resolution	Funding Mechanism(s)	Notes
Tropical Ecology Assessment and Monitoring (TEAM) Network	Citizen/community science	https://www.wildlifeinsights.org/team-network	A voluntary, decentralized network of partners who collect primary camera trap data for biodiversity monitoring that are aggregated into a Wildlife Picture Index (WPI) using AI and cloud computing. Habitat: terrestrial.	1990–present	Instantaneous	Global	Point location WPI data can be aggregated at the level of a site, sites within a region, sites within a continent, or globally	Conservation International, the Wildlife Conservation Society and the Smithsonian Institute	
Drought Network (Drought-Net)	Experimental	https://droughtnet.weebly.com/	Network of coordinated drought experiments (International Drought Experiment [IDE]), and network of existing precipitation/drought experiments (Enhancing Existing Experiments [EEE]). Habitat: terrestrial.		Milliseconds (sensors) to annual	Global	Plot/reach/transect, site	NSF	

Name	Type	URL	Description	Dates	Temporal scale	Spatial scale	Experimental unit	Funding	Notes
Nutrient Network (NutNet)	Experimental	https://nutnet.org	Grassland experiments across continents with consistent biotic and abiotic treatments (fencing, nutrients) and measuring vegetation, arthropods, soils, biogeochemistry. Habitat: terrestrial.	2007–present	Within-season to decadal	Global	1 m^2 sub-subplots within 2.5 m × 2.5 m subplots; 5 m × 5 m total experimental unit	Federal & university, institute funding	Contains observational study also; beginning in 2022: add-on disturbance and resource gradient study: DRAGNet.
Ameriflux	Experimental and observational	https://ameriflux.lbl.gov/	Abiotic and biotic (fluxes of energy, CO_2, and water vapor, methane, nitrous oxide; vegetation types, and responses). Habitat: terrestrial.	1996–present	Milliseconds to annual	Western Hemisphere	Fluxes at 30, 120, and 400 m aboveground, with flux footprints that extend out kilometers	U.S. Department of Energy, NASA, National Oceanic and Atmospheric Administration and U.S. Forest Service	ties in with the FLUXNET network
International Tundra Experiment (ITEX)	Experimental and observational	https://www.gvsu.edu/itex/	Experiments testing biotic and abiotic responses to warming through passive open top chambers in situ at tundra and some alpine sites. Habitat: terrestrial.	1991–present	Minutes to year	Circumpolar (Northern Hemisphere) + a few alpine sites	1-m^2 plots within treatment (hexagonal International Tundra Experiment open-top chambers) or ambient	Danish Polar Center and individual principal investigator funding	

continued

TABLE 4-1 Continued

Network Name	Primary Network Type	Website	Data Description	Temporal Extent	Temporal Resolution	Spatial Extent	Spatial Resolution	Funding Mechanism(s)	Notes
Detrital Input and Removal Treatment (DIRT) Network	Experimental and observational	https://dirtnet.wordpress.com/	Litter removal, addition, and controls examine how litter inputs affect soil organic matter	1956–present	Semi-decadal to decadal	Global in closed canopy, mesic forests.	Plot/Reach/Transect, Site		
TreeDivNet	Experimental and observational	https://treedivnet.ugent.be/	Over 30 sites conducting tree diversity experimental plantings and measuring tree growth and linkages between biodiversity and ecosystem function. Habitat: terrestrial.	1999–present	Annual surveys at all sites, monthly and more frequent at some	Global	16 m^2 to 400 m^2 plots in sub-hectare to many-hectare experimental platforms	Long Term Ecological Research Network (LTER), EU, various academic institutions	
BromeCast	Experimental and observational	https://bromecast.wixsite.com/home	More than 40 sites in 8 western states collect genetic, phenotypic, and demographic data on the invasive annual grass cheatgrass (*Bromus tectorum*); some sites have common gardens. Habitat: terrestrial.	2020–present	Seasonal to annual	Western North America	Sub-meter plot/reach/transect, up to site	Cooperative among universities, NGOs, and government research and land-management agencies in the United States and Canada	

Name	Type	URL	Description	Dates	Temporal resolution	Spatial extent	Spatial resolution	Funding/sponsor	Question-driven data collection
Long-Term Ecological Research Network (LTER)	Experimental and Observational	https://lternet.edu	Biotic and abiotic: meteorological, soil, organismal, biogeochemical, remote sensing data. Habitats: terrestrial, freshwater, marine.	1980–present	Milliseconds to annual	Continental U.S. & territories	Plot/reach/transect, site	NSF	
USDA Long-term Agroecosystem Research (LTAR)	Experimental and Observational	https://ltar.ars.usda.gov/	18 sites using coordinated experimentation and measurements on major U.S. agricultural ecosystems. Habitat: terrestrial.	Network: 2011–present Sites: 1910–present	Milliseconds to annual	Continental US	Sub-meter plot/reach/transect, up to site	Cooperative among USDA Agricultural Research Service, universities, and private research institutions	
USDA Experimental Forests and Ranges (EFR)	Experimental and Observational	https://www.fs.usda.gov/research/forestsandranges	81 sites across U.S., some of the longest-running, ecological experiments in the US, and many have climate, hydrologic, and ecological data spanning 100+ years. Habitat: terrestrial.	Early 1900s–present	Milliseconds to annual	Local watersheds or ranges	Sub-meter plot/reach/transect, up to site	USDA Forest Service	
National Atmospheric Deposition Program, National Trends Network	Observational	https://nadp.slh.wisc.edu/networks/national-trends-network/	271 sites that analyze inorganic forms of sulfate, nitrate, ammonium, base cations, pH, and orthophosphate (as a tracer for contamination) in precipitation. Habitats: terrestrial, freshwater.	1978–present	Weekly	Continental	Local, mostly rural sites	Cooperative among governmental agencies, educational institutions, private companies, and NGOs	

continued

TABLE 4-1 Continued

Network Name	Primary Network Type	Website	Data Description	Temporal Extent	Temporal Resolution	Spatial Extent	Spatial Resolution	Funding Mechanism(s)	Notes
National Ecological Observatory Network (NEON)	Observational	https://www.neonscience.org	Biotic and abiotic across 4 themes: Atmosphere; Biogeochemistry; Ecohydrology; Land Use, Land Cover, and Land Processes; and Organisms, Populations, and Communities (meteorological, soil, organismal, biogeochemical, and remote sensing data). Habitats: terrestrial, freshwater.	2014–present	Milliseconds to annual	Continental	Sub-meter plot/reach/transect, up to site	NSF	Hierarchical design (plots/reaches nested within sites nested within domain); researcher add-ons with assignable assets; also, relocatable units to expand data collection to new sites.
USDA Forest Inventory & Analysis (FIA)	Observational	https://www.fia.fs.usda.gov	Biotic across forest (vegetation, lichen, downed woody debris plots) and abiotic (site weather information, soil samples). Habitat: terrestrial.	1928–present	5–10 years	Continental U.S. & territories	2.07-m radius microplot; 7.3-m radius subplot; 17.95-m radius macroplot	USDA Forest Service	Standardized plots (with subplot, microplot, annual plot) revisited every 5–10 years on rotating basis;

						1998 Farm Bill designated plot data collected annually within each state.
Global Lake Ecological Observatory Network (GLEON)	Observational	https://gleon.org	Biotic and abiotic (biogeochemistry, ecohydrology, organism) Habitat: freshwater.	2005–present	Milliseconds to annual	Global	NSF Research Coordination Networks; Gordon & Betty Moore Foundation, Cary Institute for Ecosystem Studies
International Soil Carbon Network (ISCN)	Experimental and Observational	http://iscn.fluxdata.org/	Climate and geologic, soil profiles, horizon thickness, % carbon, bulk density, radiocarbon isotope data. Habitat: terrestrial.	2009–present	Milliseconds to millennia	Sub-meter to regional	USDA Forest Service, cooperative funding among multinational, private, and governmental entities

continued

TABLE 4-1 Continued

Network Name	Primary Network Type	Website	Data Description	Temporal Extent	Temporal Resolution	Spatial Extent	Spatial Resolution	Funding Mechanism(s)	Notes
GlobAllomeTree	Experimental and Observational	http://globallometree.org/	Allometric equations, tree volume and biomass, wood density, and biomass expansion factors. Habitat: terrestrial.	Network: 2013–present	Annual, decadal, centennial	Global	Plot/reach/transect, site	Food and Agriculture Organization of the United Nations (FAO), the Centre de Coopération Internationale en Recherche Agronomique pour le Développement (CIRAD), and the Department for Innovation in Biological, Agro-Food and Forest System at Tuscia University (UNITUS-DIBAF)	
National Phenology Network (NPN) and Nature's Notebook	Observational and Citizen Science	https://www.usanpn.org/	Plant and animal phenophase records. Habitats: terrestrial, coastal marine, freshwater.	2007–present	Instantaneous	Continental US	Point location data	USGS, University of Arizona, USFWS, NSF, NASA, USDA	

| Critical Zone Collaborative Network (CZ Net) and Critical Zone Exploration Network (CZEN) | Observational | https://criticalzone.org/ https://www.czen.org/ | Abiotic and biotic measurements of the environment where rock, soil, water, air, and living organisms interact and shape Earth's surface. Habitats: terrestrial, freshwater, coastal marine. | 2007–present (Critical Zone Observatories) | Various | Continental US and Puerto Rico (CZO); global (international affiliates) | Various | NSF, other (international affiliates) |

> **BOX 4-4**
> **The Nutrient Network**
>
> Ecosystems are under continual stress from a large number of human activities. Fossil fuel production and agriculture have altered global nutrient budgets, and numerous other human activities, including hunting, habitat destruction, and the import of invasive species have changed the consumers in ecosystems across the globe. The Nutrient Network (i.e., NutNet) was established to fill the widely recognized need for globally coordinated experiments to understand the impacts of human-driven changes in nutrient budgets and consumers on ecosystem dynamics. Thus, NutNet grew from a handful of sites to an expansive, integrated network of over 130 grassland sites across the globe. This growth reflects the essence of CSB—to derive insights from vast, varied terrains and interlink data across disparate ecosystems.
>
> NutNet is focused around five themes:
>
> **Productivity–Diversity Relationship:** NutNet seeks to decode the universal applicability of our current understanding of the relationship between productivity and diversity. In the context of CSB, this pertains to discerning patterns and connections spanning diverse continental ecosystems, from deserts arid grasslands to the icy realms of arctic tundra.
>
> **Nutrient Limitation Dynamics:** Delving into the co-limitation of plant production and diversity by multiple nutrients, NutNet's inquiries resonate with CSB's potential to uncover the intricate, large-scale nutrient dynamics that shape continental biomes.
>
> **Grazers, Fertilization, and Plant Ecology:** Investigating the conditions under which grazers or fertilization dictate plant biomass, diversity, and composition, NutNet's research provides insights into large-scale trophic interactions and their implications for CSB.
>
> **Consistent Data Collection Across the Globe:** In alignment with the principles of CSB, NutNet prioritizes the consistent acquisition of data across its myriad sites. This ensures direct comparability, fostering a comprehensive, continental understanding of environment-productivity–diversity relationships.

also be developed in response to requests for proposals, such as those from DOE. In the context of studying how physical, biological, and chemical processes change across scales in response to climate and anthropogenic drivers, RCNs can play a pivotal role in bringing together scientists to generate data, test hypotheses, and develop models.

Two networks have been instrumental in developing and enabling CSB—the Long Term Ecological Research (LTER) and NEON networks. The LTER network, founded in 1980, is a research network whereby scientists design and carry out long-term experimental and observational studies at one or more LTER sites to address ecological questions, often in a hypothesis-driven approach. The current 28 LTER sites span all major ecosystems across the United States and its territories, including marine, terrestrial,

Low-Investment, High-Impact Cross-Site Experimentation: Emphasizing the ethos of collective effort in CSB, NutNet promotes experiments that require minimal resource investment from individual investigators but collectively yield insights spanning diverse herbaceous ecosystems. Adopting an inclusive approach, NutNet's membership is open to ecologists dedicated to its overarching goals.

NutNet has made important discoveries about how biodiversity and ecosystem function are linked in naturally assembled ecosystems, how consumers influence ecosystem productivity, and the extent of variation in drivers of species invasion.

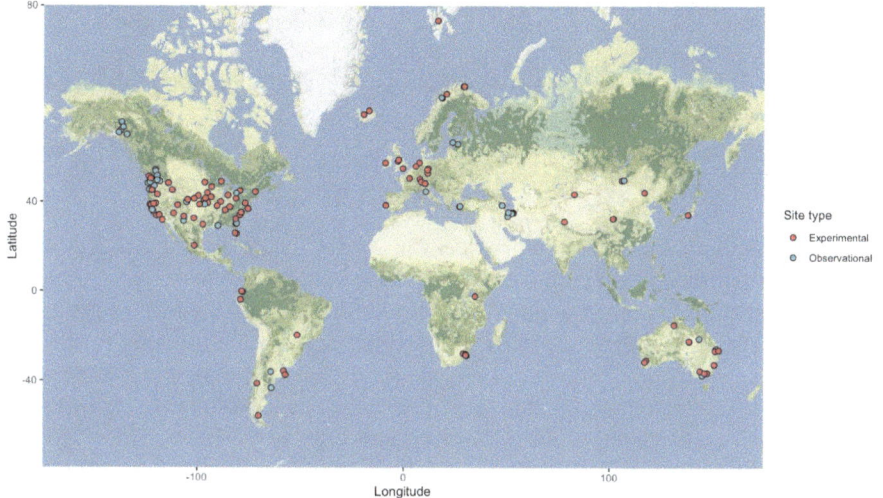

FIGURE 4-4-1 Nutrient Network site locations.
SOURCE: Elizabeth Borer and Ingrid Slette.

and freshwater sites on the North American continent, South Pacific, and Antarctica. The LTER network was instrumental in informing the design of NEON, which grew out of a desire to create coordinated and systematic observations of biotic and abiotic phenomena across ecosystems at a continental scale. This systematic approach would reduce bias and uncertainty associated with individualized measurements at many different sites, from many different studies, and where data integration and synthesis are more difficult due to the variety of approaches, instruments, and survey designs deployed at each site and even within sites.

NEON has 81 sites across the United States that are highly coordinated in a hierarchical design (Figure 4-2). Measurements and observations are systematically coordinated with standardized methods at every site. NEON is solely focused on observational approaches to measure biotic and abiotic ecological variables in a spatially nested design where plots are nested within sites, nested with 20 ecoclimate domains. All data are collected in a coordinated and standardized way within 47 terrestrial and 34 freshwater systems across the following data themes: atmosphere, biogeochemistry, ecohydrology, land cover and processes, and organisms, populations, and communities. Although separate networks, LTER and NEON are complementary; LTER's long-term record and experimental insights can be paired with NEON's spatially nested design of plots within sites within domains. In fact, 12 NEON sites are co-located with LTER sites, which directly leverage the long-term data and findings of the LTER sites. As with individual LTER sites and the NEON network, which is funded for 30 years, the longevity of a network can be finite, particularly when funding support or mandates disappear, or when questions are fully answered or change. A network and its products can then be repurposed to allow synthesis.

CENTERS OF SYNTHESIS

While collaborative networks provide a powerful means to bring together scientists across disciplines to generate and synthesize complex data across space and time, synthesis centers play an increasingly critical role in data integration, processing, and synthesis. Synthesis centers are places where researchers come together to synthesize large datasets, test hypotheses across systems, develop new theories, refine conceptual frameworks, or generate insights that pertain to large-scale biological processes, combining perspectives across disciplines. Such centers often provide resources, infrastructure, and collaborative environments to foster integrative research, especially in fields that deal with big data and complex systems, such as CSB. Such centers are generally focused on using existing data to address questions of critical importance.

Similar to the evolution of NASA missions and the evolution of networks, centers of synthesis have evolved in ways that support and enable CSB. Synthesis centers in the United States began in 1995 with the launch of the National Center for Ecological Analysis and Synthesis (NCEAS). Funded by NSF for 17 years, NCEAS focused on synthesis, data access, and collaboration. NCEAS is now funded by a range of foundations, institutes, individuals, and partnerships, and continues to host many synthesis and workshop events. Further, NCEAS now operates and hosts the LTER Network Office, enabling further synthesis within LTER. NCEAS also led to the development of the Knowledge Network for Biocomplexity Data Repository and the Environmental Data Initiative, which now connect and support data synthesis across NEON and LTER, as well as many other networks. As one of the first NSF-funded synthesis centers, the NCEAS model has led to over 20 other synthesis centers. NSF-funded synthesis centers in the United States include the Socio-Environmental Synthesis Center (SESYNC), the National Institute for Mathematical and Biological Synthesis, the National Evolutionary Synthesis Center, and now the Environmental Data Science Innovation & Inclusion Lab (ESIIL). SESYNC is aimed at bringing together natural and socioeconomic data to address questions of societal importance. ESIIL is the newest NSF–funded synthesis center aimed at developing a user community to leverage the wealth of environmental data to address environmental science questions. ESIIL emphasizes inclusion and diversity in team science and specializing in collaborative cyberinfrastructure to support the use of continental-scale data including that generated by NEON. ESIIL is explicitly envisioned as a critical synthesis center for supporting large-scale research that spans levels of biological organization, that is, described herein as CSB.

ATTRIBUTES OF SUCCESSFUL TOOLS AND NETWORKS AND CHALLENGES

Tools

Successful tools for CSB should be able to help bridge or scale processes across space and time. Most of the tools presented above allow inference across space or time and scaling or integrating across levels of biological organization. The NSF Biology Integration Institutes (BII) were established by the NSF Division of Biological Infrastructure to integrate across spatial-, temporal-, and biological scale interdisciplinary science and to promote transdisciplinary team science. The BII program specifically addresses the problem of fragmentation of the biological sciences into subdisciplines and promotes bridging scales. For example, the ASCEND (Advancing Spectral biology in Change ENvironments to understand Diversity) BII uses spectroscopic reflectance data as a common data type across scales—measured from hand-held instruments on plant tissues and from airborne and spaceborne platforms on canopies, ecosystems, and landscapes—to detect changes in biological variation and its emergent properties within whole plants, ecosystems, landscapes, and at continental scale (Box 4-3). These and related efforts have used ML and statistical models from spectroscopic imagery to predict aboveground functional traits of vegetation (Asner and Martin 2008, Miraglio et al. 2023, Wang et al. 2020), plant diversity (Gholizadeh et al. 2022, Laliberte et al. 2020, Wang et al. 2018), phylogenetic composition (Griffith et al. 2023), forest productivity (Williams et al. 2020), disease impacts on trees (Guzmán et al. 2023, Sapes et al. 2024), as well as belowground properties and microbial processes based on mechanistic linkages between aboveground and belowground portions of ecosystems (Cavender-Bares et al. 2022, Sousa et al. 2021). The integration of spectral reflectance and lidar tools with experimental, ground-based observations and laboratory analyses to test hypotheses within conceptual and theoretical frameworks is a promising approach to CSB (Box 4-3).

As the realm of CSB research continues to expand, we recognize the emergent and multifaceted challenges tied to the storage and management of biological data. The rich tapestry of life on Earth, replete with its vast array of specimens, necessitates a profound evolution in our storage methodologies. Traditionally, institutions such as museums, herbariums, and biological repositories have undertaken the formidable task of physically preserving specimens. For example, the Smithsonian National Museum of

Natural History safeguards over 148 million specimens, spanning algae to mammals[5]; the Microbiota Vault (Bello et al. 2018) is storing physical materials from global ecosystems to preserve microbial diversity. The intricacies of storing such diverse specimens demand environments with meticulous control measures to deter degradation, underlined by state-of-the-art infrastructure and consistent upkeep.

The preservation paradigm has broadened to include the digital sphere. Beyond the conventional realm of photographs and descriptive annotations, there is an influx of high-resolution 3D scans, intricate genetic sequences, and expansive data encompassing an organism's habitat and behavioral traits. Online platforms such as the Barcode of Life Data System support these diverse data types, amalgamating both molecular and morphological data, subsequently assisting researchers in their endeavors related to biodiversity and species identification. The rapid expansion of such platforms underscores an acute surge in data storage requisites. Furthermore, ensuring the accuracy of geospatial data remains paramount. Precise location details of specimens enrich biological studies, particularly those anchored in conservation, biogeography, and ecology. By facilitating the tracking of migratory patterns and assessing the repercussions of climate change on specific habitats, accurate geospatial data prove invaluable. Repositories such as the GBIF epitomize this endeavor, curating biodiversity records while preserving the integrity of geospatial information, thereby offering a comprehensive vista of biodiversity distribution. Yet, with data accumulation comes the imperative for ensuring privacy. Coupled with this is the imperative for rigorous security protocols, ensuring the safeguarding of data against potential breaches. The protocols embraced by entities such as GenBank serve as benchmarks in this domain, emphasizing data integrity and provenance and the privacy rights intrinsically linked to it. Last, the concept of "ownership" of data might prove restrictive and even inappropriate. Given the sheer scale of CSB, collaborations are inevitable. These collaborations must be based on mutual respect, trust, and a shared vision.

Networks

Attributes of networks and their "high-performing collaborative research teams" (Cheruvelil et al. 2014) that make them successful for tackling CSB include having sites within the network that represent the array of biotic and abiotic conditions across the network, and sites that have stable and adequate financial resources. Successful networks also adopt FAIR data principles, which aim to make data Findable, Accessible, Interoperable, and strive to have data Reused in other studies (Wilkinson et al. 2016). Published data that are submitted to archives, data repositories, or journals are typically assigned a Digital Object Identifier (DOI), which allows data to be found. Published data packages that contain metadata, raw (Level 0, or L0) and processed (Level 1-4, or L1-L4) data files, and any accompanying information, such as maps, documentation, or code, allow data to be reused. Interoperable data are those from

---

[5] See https://naturalhistory.si.edu/about#:~:text=We%20steward%20a%20collection%20of,moments%20we%20find%20Earth's%20story (accessed April 28, 2024).

different methods or sources that can be merged or integrated with minimal effort, for example by using standards. Efforts by iDigBio to develop a national infrastructure that oversees implementation of standards and best practices for digitization, include developing a customized cloud computing environment for collections and planning for long-term sustainability of the national digitization effort. Successful networks also adopt principles of team science, leadership, and governance (Stokols et al. 2008). They espouse inclusive and diverse membership, where members have a voice, and where team diversity (broadly defined) is effectively fostered (Cheruvelil et al. 2014). Interpersonal skills and training are taught and followed (Cheruvelil et al. 2014), and there is strong leadership from multiple people. The people within networks also share a sense of holding each other accountable to achieving the scientific goals and delivering to a diverse group of stakeholders.

Many networks formally solicit input from and transfer knowledge to their stakeholder groups. For example, the USDA Forest Service FIA program solicits and listens to user feedback through an annual user group meeting, as well as from collaborators at academic, other government, and nongovernmental organizations. FIA has adapted to changing user needs by increasing their capacity to analyze and publish data, and by expanding the scope of their data collection to include an additional suite of attributes on a subsample of plots such as soils, understory vegetation, tree crown conditions, down woody material, and invasive species.

Networks also face numerous challenges, from data inoperability to funding. Some networks are comprised of sites that measure the same process but use widely differing methods to do so. For example, the LTER network is comprised of separate sites that conduct long-term, place-based observations. Measurement approaches and frequencies are not standardized across sites. Thus, comparing or synthesizing observations across the network or subsets of sites is challenging. NEON attempted to address the limitations of the LTERs by collecting systematic measurements across sites, but NEON data are big and complex. NEON sites collect 110 TB yr^{-1}, and many NEON data products are not model-ready. Raw sensor or human-collected data (L0) are available upon request, but progressively processed data (e.g., L1–L4) are served on the NEON Data Portal and readily usable. One challenge for NEON is that NEON data have a high level of uncertainty at the continental scale; more data coverage across more ecosystems is needed to be of most use to CSB.

Even if network data are readily available and usable, there remains a challenge of data "unfamiliarity," or data users who are one step removed from the physical data collection and thus unfamiliar with the data's limitations. Historically, this challenge is overcome by involving the principal investigator (PI) in data analysis and/or publications. But, widening a research team takes time and effort to build relationships, and if involving the PI is not an option, then data usage and scientific advancements can be limited. One network is experimenting with this challenge. Data collected across the Ameriflux network previously fell under their "legacy data policy," which mandated that the PI be a co-author. Ameriflux now has an open-data policy, where the PI does not have to be on the publication. This experiment could be risky, because it removes the site knowledge from the paper.

Last, maintaining continuity and integrity of research networks is a challenge. With gaps in funding or the end of funding, maintaining a data stream may not be possible and thus especially disruptive to long-term research. When network sites, campaigns, or instruments are decommissioned due to unforeseen or planned circumstances, a break in data collection and availability can have large impacts on the inferences made from the research involving those data. Especially for networks that contain numerous individual research sites and data streams, researchers need resources and guidance on how to avoid or minimize the gap in data availability. NSF's data management plans require investigators to address the physical and/or cyber resources and facilities that will be used to store and preserve the data after the grant ends, yet there is a lack of recommendations on how to minimize negative impacts of gaps in data collection due to sites or instruments temporarily or permanently dropping out of a network. There is also no standard mechanism for "emergency" funds that would help minimize the impacts. For example, LTER was established in 1980, and sites are renewed through NSF's LTER program. Originally, six LTER sites were established (Andrews Forest, Coweeta, Konza Prairie, Niwot Ridge, North Temperate Lakes, North Inlet). Of these, all but North Inlet and Coweeta have continued. Since the initial sites were established, the LTER program has expanded to include as many as 32 sites. Over time, some sites have not been renewed through the renewal process every 5 years. In most cases, these sites are co-located with existing biological stations with other research activities ongoing. The use of LTER-funded equipment therefore lives on past the lifespan of the LTER. However, there may not be sufficient support to continue data collection, leading to a break in the long-term data. This is problematic because gaps in data may miss important interannual variability and alter conclusions about long-term trends.

RECOMMENDATION AND CONCLUSION

Fully responding to the challenges of developing CSB requires both the enhancement of existing and the development of new infrastructure. The committee offers the following recommendation and conclusion to NSF and other agencies.

Recommendation 4-1: To provide infrastructure specifically aimed at supporting continental-scale biology (CSB), the National Science Foundation (NSF) should consider the following options, as available resources permit.
- Explore the development of artificial intelligence (AI) and informatics tools, and open-access databases explicitly focused on CSB, synthesizing knowledge across scales, that would synergize with the synthesis work currently conducted at the Environmental Data Science Innovation & Inclusion Lab. A request for proposals (RFP) to support virtual infrastructure and computational science innovations would be of great value to realize the potential of CSB data. For example, linking remotely sensed spectral and structural measurements to physical and biological measurements on the ground could be advanced by developing new algorithmic modalities.
- Build new sensor modalities to improve data collection. This could be achieved by developing interdisciplinary funding opportunities that unite

ecology, engineering, atmospheric science, remote sensing, hydrology, and other disciplines such as social sciences.
- Allocate resources for next-generation digitization of biodiversity collections to enhance their utility as reference standards for CSB and to enable the development of digital ecosystem twins. This will require new bioinformatics tools to enable access to and management of preserved and living collections to facilitate their utility for interpretation of in situ and remotely sensed data.
- Develop communities that can leverage interdisciplinary data from NSF platforms and various networks, akin to the use of National Ecological Observatory Network (NEON) data by researchers funded by the previous Macrosystem Biology Program. An example would be incorporating macrosystems/synthesis research to create living data products (those that are continually updated) that inform biological processes at continental scale. If done, data from one platform could serve as calibration/validation for other data products and layers from other platforms to facilitate interpolation, extrapolation, and/or imputation. Major government assessments, such as the National Nature Assessment, could also leverage data provided by integration of these platforms.
- Explore joint support of integrative science via interagency (e.g., NASA, U.S. Department of Energy) RFPs, for example, multiscale coordinated interdisciplinary field campaigns (e.g., Arctic-Boreal Vulnerability Experiment, the Biodiversity Survey of the Cape, and the Large-scale Biosphere-Atmosphere Experiment in Amazonia).
- Support efforts to understand how to sample for continental-scale biological questions. Few spatially distributed networks with standardized sampling exist, and those that do exist require great resources. Investment in research to understand sampling theory (time and space) for capturing continental scales and cross-boundary interactions (e.g., metacoupling, as discussed in Chapter 2) is needed.
- Explore development of interagency incentives and mechanisms for public–private partnerships that can facilitate targeted private investment in data development and integration across scales and types, for example, engaging in AI-driven data analysis and data product development.

Conclusion 4-1: Development of research infrastructure for CSB would also benefit from actions by other agencies. Examples include the following:
- The Small Business Administration could develop agency-specific innovation research (Small Business Innovation Research) and technology transfer research (Small Business Technology Transfer Research) RFPs that focus on AI, machine learning, and sensor development for the biological and environmental sciences.
- NASA's continued support for the Surface Biology and Geology (SBG) mission, the only satellite instrument dedicated to biological processes that will specifically enable CSB, is an important contribution. SBG will provide

continuous data across continents and the globe to fill in the gaps from NEON and enable baseline information to track changes in biological processes through time. Other NASA Explorer and Incubator missions recommended by the Decadal Survey in Earth Science and Applications from Space (NASEM 2018) complement SBG via lidar and radar measurements of ecosystem three-dimensional D structure.

- Agencies engaged in these efforts could continue to support scientific assessments and action-oriented efforts that inform policies guided by or consistent with the UN Convention on Biological Diversity. These include the Global Biodiversity Information Facility, the Group on Earth Observations Biodiversity Observation Network, the Intergovernmental Science-Policy Platform on Biodiversity and Ecosystem Services.
- Continue support of related domestic efforts that contribute to CSB, for example, the ongoing National Nature Assessment, U.S. Geological Survey national Biodiversity and Climate Change Assessment, and the U.S. contribution to the 30×30 Conservation initiative (White House, 2021).

REFERENCES

Asner, G.P., and R.E. Martin. 2008. Spectral and chemical analysis of tropical forests: Scaling from leaf to canopy levels. *Remote Sensing of Environment* 112:3958-3970. https://doi.org/10.1016/j.rse.2008.07.003.

Avissar, R., and C.A. Nobre. 2002. Preface to special issue on the Large-Scale Biosphere-Atmosphere Experiment in Amazonia (LBA). *Journal of Geophysical Research: Atmospheres* 107(D20):8034. https://doi.org/10.1029/2002JD002507.

Avissar, R., P.L. Silva Dias, M.A. Silva Dias, and C. Nobre. 2002. The Large-Scale Biosphere-Atmosphere Experiment in Amazonia (LBA): Insights and future research needs. *Journal of Geophysical Research: Atmospheres* 107(D20):8086. https://doi.org/10.1029/2002JD002704.

Barberán, A., K.L. McGuire, J.A. Wolf, F.A. Jones, S.J. Wright, B.L. Turner, A. Essene, S.P. Hubbell, B.C. Faircloth, and N. Fierer. 2015. Relating belowground microbial composition to the taxonomic, phylogenetic, and functional trait distributions of trees in a tropical forest. *Ecology Letters* 18:1397-1405. https://doi.org/10.1111/ele.12536.

Bello, M.G.D., R. Knight, J.A. Gilbert, and M.J. Blaser. 2018. Preserving microbial diversity. *Science* 362:33-34. https://doi.org/10.1126/science.aau8816.

Berner, L.T., and S.J. Goetz, S.J. 2022. Satellite observations document trends consistent with a boreal forest biome shift. *Global Change Biology* 28:3275-3292. https://doi.org/10.1111/gcb.16121.

Blondeel, H., J. Guillemot, N. Martin-StPaul, A. Druel, S. Bilodeau-Gauthier, J. Bauhus, C. Grossiord, A. Hector, H. Jactel, J. Jensen, C. Messier, B. Muys, H. Serrano-León, et al. 2024. Tree diversity reduces variability in sapling survival under drought. *Journal of Ecology* 112(5):1164-1180. https://doi.org/10.1111/1365-2745.14294.

Borton, M.A., B.B. McGivern, K.R. Willi, B.J. Woodcroft, A.C. Mosier, D.M. Singleton, T. Bambakidis, et al. 2023. A functional microbiome catalog crowdsourced from North American rivers. *bioRxiv*. https://doi.org/10.1101/2023.07.22.550117.

Brewer, T., K. Handley, P. Carini, J.A. Gilbert, and N. Fierer. 2017. Genome reduction in an abundant and ubiquitous soil bacterium "*Candidatus* Udaeobacter copiosus." *Nature Microbiology* 2:16198. https://doi.org/10.1038/nmicrobiol.2016.198.

Buzzard, V., S.T. Michaletz, Y. Deng, Z. He, D. Ning, L. Shen, Q. Tu, et al. 2019. Continental scale structuring of forest and soil diversity via functional traits. *Nature Ecology & Evolution* 3:1298-1308. https://doi.org/10.1038/s41559-019-0954-7.

Cavender-Bares, J., A.K. Schweiger, J.N. Pinto-Ledezma, and J.E. Meireles. 2020. Applying remote sensing to biodiversity science. Pp. 13-42 in *Remote Sensing of Plant Biodiversity*, J. Cavender-Bares, J.A. Gamon, and P.A. Townsend, eds. Cham, Switzerland: Springer International.

Cavender-Bares, J., A.K. Schweiger, J.A. Gamon, H. Gholizadeh, K. Helzer, C. Lapadat, M.D. Madritch, P.A. Townsend, Z. Wang, and S.E. Hobbie. 2022. Remotely detected aboveground plant function predicts belowground processes in two prairie diversity experiments. *Ecological Monographs* 92(1):e01488. https://doi.org/10.1002/ecm.1488.

Cheruvelil, K.S., P.A. Soranno, K.C. Weathers, P.C. Hanson, S.J. Goring, C.T. Filstrup, and E.K. Read. 2014. Creating and maintaining high-performing collaborative research teams: The importance of diversity and interpersonal skills. *Frontiers in Ecology and the Environment* 12:31-38. https://doi.org/10.1890/130001.

Chylek, P., C. Folland, J.D. Klett, M. Wang, N. Hengartner, G. Lesins, and M.K. Dubey. 2022. Annual mean arctic amplification 1970–2020: Observed and simulated by CMIP6 climate models. *Geophysical Research Letters* 49(13):e2022GL099371. https://doi.org/10.1029/2022GL099371.

Clark, M.L., L. Salas, S. Baligar, C.A. Quinn, R.L. Snyder, D. Leland, W. Schackwitz, S.J. Goetz, and S. Newsam. 2023. The effect of soundscape composition on bird vocalization classification in a citizen science biodiversity monitoring project. *Ecological Informatics* 75:102065. https://doi.org/10.1016/j.ecoinf.2023.102065.

Davidson, S.C., G. Bohrer, E. Gurarie, S. LaPoint, P.J. Mahoney, N.T. Boelman, J.U.H. Eitel, L.R. Prugh, L.A. Vierling, J. Jennewein, E. Grier, O. Couriot, A.P. Kelly, A.J.H. Meddens, R.Y. Oliver, R. Kays, M. Wikelski, et al. 2020. Ecological insights from three decades of animal movement tracking across a changing Arctic. *Science* 370:712-715. https://doi.org/10.1126/science.abb7080.

de Koning, K., J. Broekhuijsen, I. Kühn, O. Ovaskainen, F. Taubert, D. Endresen, D. Schigel, and V. Grimm. 2023. Digital twins: Dynamic model-data fusion for ecology. *Trends in Ecology & Evolution* 38(10):916-926. https://doi.org/10.1016/j.tree.2023.04.010.

Dial, R.J., C.T. Maher, R.E. Hewitt, and P.F. Sullivan. 2022. Sufficient conditions for rapid range expansion of a boreal conifer. *Nature* 608:546-551. https://doi.org/10.1038/s41586-022-05093-2.

Dubayah, R., J.B. Blair, S. Goetz, L. Fatoyinbo, M. Hansen, S. Healey, M. Hofton, G. Hurtt, J. Kellner, S. Luthcke, J. Armston, H. Tang, L. Duncanson, S. Hancock, P. Jantz, S. Marselis, P.L. Patterson, W. Qi, and C. Silva. 2020. The Global Ecosystem Dynamics Investigation: High-resolution laser ranging of the Earth's forests and topography, *Science of Remote Sensing* 1:100002. https://doi.org/10.1016/j.srs.2020.100002.

Dubayah, R., M. Hofton, J.B. Blair, J. Armston, H. Tang, S. Luthcke. 2021. GEDI L2A elevation and height metrics data global footprint level V002. NASA EOSDIS Land Processes DAAC. https://doi.org/10.5067/GEDI/GEDI02_A.002.

Duncanson, L., J.R. Kellner, J. Armston, R. Dubayah, D.M. Minor, S. Hancock, S.P. Healey, P.L. Patterson, S. Saarela, S. Marselis, C.E. Silva, J. Bruening, S.J. Goetz, H. Tang, M. Hofton, B. Blair, S. Luthcke, L. Fatoyinbo, et al. 2022. Aboveground biomass density models for NASA's Global Ecosystem Dynamics Investigation (GEDI) lidar mission. *Remote Sensing of Environment* 270:112845. https://doi.org/10.1016/j.rse.2021.112845.

Enquist, B.J., R. Condit, R.K. Peet, M. Schildhauer, and B. Thiers. 2009. Cyberinfrastructure for an integrated botanical information network to investigate the ecological impacts of global climate change on plant biodiversity. http://dx.doi.org/10.7287/PEERJ.PREPRINTS.2615.

Fierer, N., J.W. Leff, B.J. Adams, U.N. Nielsen, S.T. Bates, C.L. Lauber, S. Owens, J.A. Gilbert, D.H. Wall, and J.G. Caporaso. 2012. Cross-biome metagenomic analyses of soil microbial communities and their functional attributes. *Proceedings of the National Academy of Sciences of the United States of America* 109:21390-21395. https://doi.org/10.1073/pnas.1215210110.

Fierer, N., J. Ladau, J.C. Clemente, J.W. Leff, S.M. Owens, K.S. Pollard, R. Knight, J.A. Gilbert, and R.L. Mcculley. 2013. Reconstructing the microbial diversity and function of pre-agricultural tallgrass prairie soils in the United States. *Science* 342:621-624. https://doi.org/10.1126/science.1243768.

Fontes, C.G., J. Pinto-Ledezma, A.L. Jacobsen, R.B. Pratt, and J. Cavender-Bares. 2022. Adaptive variation among oaks in wood anatomical properties is shaped by climate of origin and shows limited plasticity across environments. *Functional Ecology* 36:326-340. https://doi.org/10.1111/1365-2435.13964.

Forkel, M., N. Carvalhais, C. Rödenbeck. R. Keeling, M. Heimann, K. Thonicke, S. Zaehle, and M. Reichstein. 2016. Enhanced seasonal CO_2 exchange caused by amplified plant productivity in northern ecosystems. *Science* 351:696-699. https://doi.org/10.1126/science.aac4971.

Foster, A.C., J.A. Wang, G.V. Frost, S.J Davidson, E. Hoy, K.W. Turner, O. Sonnentag, et al. 2022. Disturbances in North American boreal forest and Arctic tundra: Impacts, interactions, and responses. *Environmental Research Letters* 17(11):113001. https://doi.org/10.1088/1748-9326/ac98d7.

Franklin, J. 2010. *Mapping Species Distributions: Spatial Inference and Prediction.* Cambridge, UK. Cambridge University Press.

Frans, V., and J. Liu. 2024. Gaps and opportunities in modeling human influence on species distributions in the Anthropocene. *Nature Ecology & Evolution* https://www.nature.com/articles/s41559-024-02435-3.

Fremin, B.J., H. Sberro, and A.S. Bhatt. 2020. MetaRibo-Seq measures translation in microbiomes. *Nature Communications* 11(1): 3268. https://doi.org/10.1038/s41467-020-17081-z.

Fretwell, P.T., and P.N. Trathan, 2021. Discovery of new colonies by Sentinel2 reveals good and bad news for emperor penguins. *Remote Sensing in Ecology and Conservation* 7:139-153. https://doi.org/10.1002/rse2.176.

Gholizadeh, H., J.A. Gamon, P.A. Townsend, A.I. Zygielbaum, C.J. Helzer, G.Y. Hmimina, R. Yu, R.M. Moore, A.K. Schweiger, and J. Cavender-Bares. 2019. Detecting prairie biodiversity with airborne remote sensing. *Remote Sensing of Environment* 221:38-49. https://doi.org/10.1016/j.rse.2018.10.037.

Gholizadeh, H., J.A. Gamon, C.J. Helzer, and J. Cavender-Bares. 2020. Multi-temporal assessment of grassland α- and β-diversity using hyperspectral imaging. *Ecological Applications* 30:e02145. https://doi.org/10.1002/eap.2145.

Gholizadeh, H., A.P. Dixon, K.H. Pan, N.A. McMillan, R.G. Hamilton, S.D. Fuhlendorf, J. Cavender-Bares, and J.A. Gamon. 2022. Using airborne and DESIS imaging spectroscopy to map plant diversity across the largest contiguous tract of tallgrass prairie on Earth. *Remote Sensing of Environment* 281:113254. https://doi.org/10.1016/j.rse.2022.113254.

Gibbons, S.M., J.G. Caporaso, M. Pirrung, D. Field, R. Knight, and J.A. Gilbert. 2013. Evidence for a persistent microbial seed bank throughout the global ocean. *Proceedings of the National Academy of Sciences of the United States of America* 110:4651-4655. https://doi.org/10.1073/pnas.1217767110.

Gilbert, J.A., J.A. Steele, J.G. Caporaso, L. Steinbrück, J. Reeder, B. Temperton, S. Huse, A.C. McHardy, R. Knight, I. Joint, P. Somerfield, J.A. Fuhrman, and D. Field. 2012. Defining seasonal marine microbial community dynamics. The *ISME Journal* 6(2):298-308. https://doi.org/10.1038/ismej.2011.107.

Graven, H.D. 2015. Impact of fossil fuel emissions on atmospheric radiocarbon and various applications of radiocarbon over this century. *Proceedings of the National Academy of Sciences of the United States of America* 112:9542-9545. https://doi.org/10.1073/pnas.1504467112.

Grieneisen, L., M. Dasari, T.J. Gould, J.R. Björk, J.-C. Grenier, V. Yotova, D. Jansen, et al. 2021. Gut microbiome heritability is nearly universal but environmentally contingent. *Science* 373:181-186. https://doi.org/10.1126/science.aba5483.

Griffith, D.M., K.B. Byrd, L.D.L. Anderegg, E. Allan, D. Gatziolis, D. Roberts, R. Yacoub, and R.R. Nemani. 2023. Capturing patterns of evolutionary relatedness with reflectance spectra to model and monitor biodiversity. *Proceedings of the National Academy of Sciences of the United States of America* 120(24):e2215533120. https://doi.org/10.1073/pnas.2215533120.

Guzmán, Q.J.A., J.N. Pinto-Ledezma, D. Frantz, P.A. Townsend, J. Juzwik and J. Cavender-Bares. 2023. Mapping oak wilt disease from space using land surface phenology. *Remote Sensing of Environment* 298:113794. https://doi.org/10.1016/j.rse.2023.113794.

Hagedorn, F., J. Joseph, M. Peter, J. Luster, K. Pritsch, U. Geppert, R. Kerner, V. Molinier, S. Egli, M. Schaub, J.-F. Liu, M. Li, K. Sever, M. Weiler, R.T.W. Siegwolf, A. Gessler, and M. Arend. 2016. Recovery of trees from drought depends on belowground sink control. *Nature Plants* 2:1-5. https://doi.org/10.1038/nplants.2016.111.

Hakkenberg, C.R., J.W. Atkins, J.F. Brodie, P. Burns, S. Cushman, P. Jantz, Z. Kaszta, C.A. Quinn, M.D. Rose, and S.J. Goetz. 2023. Inferring alpha, beta, and gamma plant diversity across biomes with GEDI spaceborne lidar. *Environmental Research: Ecology* 2(3):035005. https://doi.org/10.1088/2752-664X/acffcd.

Hansen, M.C., P.V. Potapov, R. Moore, M. Hancher, S.A. Turubanova, A. Tyukavina, D. Thau, S.V. Stehman, S.J. Goetz, T.R. Loveland, A. Kommareddy, A. Egorov, L. Chini, C.O. Justice, and J.R.G. Townshend. 2013. High-resolution global maps of 21st-century forest cover change. *Science* 342:850-853. https://doi.org/10.1126/science.1244693.

Hasler, C. 2022. Radiocarbon's blast from the past. *Eos* 103 https://doi.org/10.1029/2022EO220445.

Isbell, F., D. Craven, J. Connolly, M. Loreau, B. Schmid, C. Beierkuhnlein, T.M. Bezemer, et al. 2015. Biodiversity increases the resistance of ecosystem productivity to climate extremes. *Nature* 526:574-577. https://doi.org/10.1038/nature15374.

Jovan, S., M. Haldeman, S. Will-Wolf, K. Dillman, L. Geiser, J. Thompson, D. Stone, and J. Hollinger. 2021. *National Atlas of Epiphytic Lichens in Forested Habitats of the United States*. Gen. Tech. Rep. PNW-GTR-986. Portland, OR: U.S. Department of Agriculture, Forest Service, Pacific Northwest Research Station.

Kahl, S., C.M. Wood, M. Eibl, and H. Klinck. 2021. BirdNET: A deep learning solution for avian diversity monitoring. *Ecological Informatics* 61:101236. https://doi.org/10.1016/j.ecoinf.2021.101236.

Kattge, J., G. Bönisch, S. Díaz, S. Lavorel, I.C. Prentice, P. Leadley, S. Tautenhahn, G.D.A. Werner, et al. 2020. TRY plant trait database—enhanced coverage and open access. *Global Change Biology* 26:119-188. https://doi.org/10.1111/gcb.14904.

Kays, R., S.C. Davidson, M. Berger, W. Fiedler, A. Flack, J. Hirt, C. Hahn, D. Gauggel, B. Russell, et al. 2022. The Movebank system for studying global animal movement and demography. *Methods in Ecology and Evolution* 13:419-431. https://doi.org/10.1111/2041-210X.13767.

Keenan, T., D. Hollinger, G. Bohrer, D. Dragoni, J.W. Munger, H.P. Schmid, and A.D. Richardson. 2013. Increase in forest water-use efficiency as atmospheric carbon dioxide concentrations rise. *Nature* 499:324-327. https://doi.org/10.1038/nature12291.

Kolton, M., J.W. David, X. Mayali, P.K. Weber, K.J. McFarlane, J. Pett-Ridge, M.M. Somoza, J. Lietard, J.B. Glass, E.A. Lilleskov, A.J. Shaw, S. Tringe, P.J. Hanson, J.E. Kostka. 2022. Defining the *Sphagnum* core microbiome across the North American continent reveals a central role for diazotrophic methanotrophs in the nitrogen and carbon cycles of boreal peatland ecosystems. *mBio* 13:e03714-03721. https://doi.org/10.1128/MBIO.03714-21.

Kothari, S., R. Beauchamp-Rioux, E. Laliberté, and J. Cavender-Bares. 2023. Reflectance spectroscopy allows rapid, accurate and non-destructive estimates of functional traits from pressed leaves. *Methods in Ecology and Evolution* 14:385-401. https://doi.org/10.1111/2041-210X.13958.

Kravitz, B., A.B. Guenther, L. Gu, T. Karl, L. Kaser, S.G. Pallardy, J. Peñuelas, M.J. Potosnak, R. Seco. 2016. A new paradigm of quantifying ecosystem stress through chemical signatures. *Ecosphere* 7(11):e01559. https://doi.org/10.1002/ecs2.1559.

Kurczy, M.E., E.M. Forsberg, M.P. Thorgersen, F.L. Poole III, H.P. Benton, J. Ivanisevic, M.L. Tran, J.D. Wall, D.A. Elias, M.W.W. Adams, and G. Siuzdak. 2016. Global isotope metabolomics reveals adaptive strategies for nitrogen assimilation. *ACS Chemical Biology* 11:1677-1685. https://doi.org/10.1021/acschembio.6b00082.

Ladau, J., Y. Shi, X. Jing, J.-S. He, L. Chen, X. Lin, N. Fierer, J.A. Gilbert, K.S. Pollard, and H. Chu. 2018. Existing climate change will lead to pronounced shifts in the diversity of soil prokaryotes. *MSystems* 3(5). https://doi.org/10.1128/msystems.00167-18.

Laliberté, E., A.K. Schweiger, and P. Legendre. 2020. Partitioning plant spectral diversity into alpha and beta components. *Ecology Letters* 23:370-380. https://doi.org/10.1111/ele.13429.

Lang, A.K., E.A. LaRue, and S.N. Kivlin. 2023. Forest structural diversity is linked to soil microbial diversity. *Ecosphere* 14(11):e4702. https://doi.org/10.1002/ecs2.4702.

Larsen, P.E., N. Scott, and A.F. Post. 2015. Satellite remote sensing data can be used to model marine microbial metabolite turnover. *The ISME Journal* 9(1):166-179. https://doi.org/10.1038/ismej.2014.107.

Liu, Y., J.A. Hogan, J.W. Lichstein, R.P. Guralnick, D.E. Soltis, P.S. Soltis, and S.M. Scheiner. 2024. Biodiversity and productivity in eastern US forests. *Proceedings of the National Academy of Sciences of the United States of America* 121(14):e2314231121. https://doi.org/10.1073/pnas.2314231121.

Long, P.E., K.H. Williams, S.S. Hubbard, and J.F. Banfield. 2016. Microbial metagenomics reveals climate-relevant subsurface biogeochemical processes. *Trends in Microbiology* 24:600-610. https://doi.org/10.1016/j.tim.2016.04.006.

Ma, L., G. Hurtt, H. Tang, R. Lamb, A. Lister, L. Chini, R. Dubayah, J. Armston, E. Campbell, L. Duncanson, S. Healey, J. O'Neil-Dunne, L. Ott, B. Poulter, and Q. Shen. 2023. Spatial heterogeneity of global forest aboveground carbon stocks and fluxes constrained by spaceborne lidar data and mechanistic modeling. *Global Change Biology* 29:3378-3394. doi.org/10.1111/gcb.16682.

Massey, R., B.M. Rogers, L.T. Berner, S, Cooperdock, M.C. Mack, X.J. Walker, and S.J. Goetz. 2023. Forest composition change and biophysical climate feedbacks across boreal North America. *Nature Climate Change* 13:1368-1375. https://doi.org/10.1038/s41558-023-01851-w.

Meireles, J.E., J. Cavender-Bares, P.A. Townsend, S. Ustin, J.A. Gamon, A.K. Schweiger, M.E. Schaepman, G.P. Asner, R.E. Martin, and A. Singh. 2020. Leaf reflectance spectra capture the evolutionary history of seed plants. *New Phytologist* 228:485-493. https://doi.org/10.1111/nph.16771.

Miniat, C.F., A.C. Oishi, P.V. Bolstad, C.R. Jackson, N. Liu, J.P. Love, C.M. Pringle, K.J. Solomon, and N. Wurzburger. 2021. The Coweeta Hydrologic Laboratory and the Coweeta Long-Term Ecological Research Project. *Hydrological Processes* 35(7):e14302. https://doi.org/10.1002/hyp.14302.

Miraglio, T., N.C. Coops, C.I.B. Wallis, A.L. Crofts, M. Kalacska, M. Vellend, S.P. Serbin, J.P. Arroyo-Mora, and E. Laliberté. 2023. Mapping canopy traits over Québec using airborne and spaceborne imaging spectroscopy. *Scientific Reports* 13:17179. https://doi.org/10.1038/s41598-023-44384-0.

NASEM (National Academies of Sciences, Engineering, and Medicine). 2018. *Thriving on Our Changing Planet: A Decadal Strategy for Earth Observation from Space*. Washington, DC: The National Academies Press. https://doi.org/10.17226/24938.

NASEM. 2020. *Biological Collections: Ensuring Critical Research and Education for the 21st Century*. Washington, DC: The National Academies Press. https://doi.org/10.17226/25592.

NASEM. 2024. *Foundational Research Gaps and Future Directions for Digital Twins*. Washington, DC: The National Academies Press. https://doi.org/10.17226/26894.

Nayfach, S., S. Roux, R. Seshadri, D. Udwary, N. Varghese, F. Schulz, D. Wu, et al. 2021. A genomic catalog of Earth's microbiomes. *Nature Biotechnology* 39:499-509. https://doi.org/10.1038/s41587-020-0718-6.

NRC (National Research Council). 2015. *Enhancing the Effectiveness of Team Science*. Washington, DC: The National Academies Press. https://doi.org/10.17226/19007.

O'Connell, A.F., J.D. Nichols, and K.U. Karanth (eds.). 2011. *Camera Traps in Animal Ecology: Methods and Analyses*. Springer.

Osburn, E.D., C.F. Miniat, K.J. Elliott, and J.E. Barrett. 2021. Effects of Rhododendron removal on soil bacterial and fungal communities in southern Appalachian forests. *Forest Ecology and Management* 496:119398. https://doi.org/10.1016/j.foreco.2021.119398.

Pastore, M.A., S.E. Hobbie, and P.B. Reich. 2021. Sensitivity of grassland carbon pools to plant diversity, elevated CO_2, and soil nitrogen addition over 19 years. *Proceedings of the National Academy of Sciences of the United States of America* 118:e2016965118. https://doi.org/10.1073%2Fpnas.2016965118.

Pellegrini, A.F.A., A.M. Hein, J. Cavender-Bares, R.A. Montgomery, A.C. Staver, F. Silla, S.E. Hobbie, and P.B. Reich. 2021. Disease and fire interact to influence transitions between savanna–forest ecosystems over a multi-decadal experiment. *Ecology Letters* 24:1007-1017. https://doi.org/10.1111/ele.13719.

Perez, T.M., O. Valverde-Barrantes, C. Bravo, T.C. Taylor, B. Fadrique, J.A. Hogan, C.J. Pardo, J.T. Stroud, C. Baraloto, and K.J. Feeley. 2019. Botanic gardens are an untapped resource for studying the functional ecology of tropical plants. *Philosophical Transactions of the Royal Society B: Biological Sciences* 374(1763):20170390. https://doi.org/10.1098/rstb.2017.0390.

Pinto-Ledezma, J.N., and J. Cavender-Bares. 2021. Predicting species distributions and community composition using satellite remote sensing predictors. *Scientific Reports* 11:16448. https://doi.org/10.1038/s41598-021-96047-7.

Poo, S., S.M. Whitfield, A. Shepack, G.J. Watkins-Colwell, G. Nelson, J. Goodwin, A. Bogisich, et al. 2022. Bridging the research gap between live collections in zoos and preserved collections in natural history museums. *BioScience* 72:449-460. https://doi.org/10.1093/biosci/biac022.

Potapov, P., M.C. Hansen, A. Pickens, A. Hernandez-Serna, A. Tyukavina, S. Turubanova, V. Zalles, X. Li, A. Khan, F. Stolle, N. Harris, X.P. Song, A. Baggett, I. Kommareddy, A. Kommareddy. 2022. The global 2000-2020 land cover and land use change dataset derived from the Landsat archive: First results. *Frontiers in Remote Sensing* 3:856903. https://doi.org/10.3389/frsen.2022.856903.

Previdi, M., K.L. Smith, and L.M. Polvani. 2021. Arctic amplification of climate change: A review of underlying mechanisms. *Environmental Research Letters* 16(9):093003. https://doi.org/10.1088/1748-9326/ac1c29.

Prosser, J.I. 2015. Dispersing misconceptions and identifying opportunities for the use of 'omics' in soil microbial ecology. *Nature Reviews Microbiology* 13:439-446. https://doi.org/10.1038/nrmicro3468.

Püttker, S., F. Kohrs, D. Benndorf, R. Heyer, E. Rapp, and U. Reichl. 2015. Metaproteomics of activated sludge from a wastewater treatment plant–A pilot study. *Proteomics* 15:3596-3601. https://doi.org/10.1002/pmic.201400559.

Quinn, C.A., P. Burns, C.R. Hakkenberg, L. Salas, B. Pasch, S.J. Goetz, and M.L. Clark. 2023. Soundscape components inform acoustic index patterns and refine estimates of bird species richness. *Frontiers in Remote Sensing* 4:1156837. https://doi.org/10.3389/frsen.2023.1156837.

Rantanen, M., A.Y. Karpechko, A. Lipponen, K. Nordling, O. Hyvärinen, K. Ruosteenoja, T. Vihma, and A. Laaksonen. 2022. The Arctic has warmed nearly four times faster than the globe since 1979. *Communications Earth & Environment* 3(1):168. https://doi.org/10.1038/s43247-022-00498-3.

Reich, P.B. 2009. Elevated CO_2 reduces losses of plant diversity caused by nitrogen deposition. *Science* 326:1399-1402. https://doi.org/10.1126/science.1178820.

Reich, P.B., K.M. Sendall, A. Stefanski, R.L. Rich, S.E. Hobbie, and R.A. Montgomery. 2018. Effects of climate warming on photosynthesis in boreal tree species depend on soil moisture. *Nature* 562:263-267. https://doi.org/10.1038/s41586-018-0582-4.

Root, H.T., L.H. Geiser, S. Jovan, and P. Neitlich. 2015. Epiphytic macrolichen indication of air quality and climate in interior forested mountains of the Pacific Northwest, USA. *Ecological Indicators* 53:95-105. https://doi.org/10.1016/j.ecolind.2015.01.029.

Ruff, Z.J., D.B. Lesmeister, J.M. Jenkins, and C.M. Sullivan. 2023. PNW-Cnet v4: Automated species identification for passive acoustic monitoring. *SoftwareX* 23:101473. https://doi.org/10.1016/j.softx.2023.101473.

Ryu, Y., J.A. Berry, and D.D. Baldocchi. 2019. What is global photosynthesis? History, uncertainties and opportunities. *Remote Sensing of Environment* 223:95-114. https://doi.org/10.1016/j.rse.2019.01.016.

Sapes, G., L. Schroeder, A. Scott, I. Clark, J. Juzwik, R.A. Montgomery, J.A. Guzmán, J. Cavender-Bares. 2024. Mechanistic links between physiology and spectral reflectance enable previsual detection of oak wilt and drought stress. *Proceedings of the National Academy of Sciences of the United States of America* 121(7):e2316164121 https://doi.org/10.1073/pnas.2316164121.

Schwager, P., and C. Berg. 2021. Remote sensing variables improve species distribution models for alpine plant species. *Basic and Applied Ecology* 54:1-13. https://doi.org/10.1016/j.baae.2021.04.002.

Sellers, P.J., F.G. Hall, G. Asrar, D.E. Strebel, and R.E. Murphy. 1988. The First ISLSCP Field Experiment (FIFE). *Bulletin of the American Meteorological Society* 69(1):22-27. http://www.jstor.org/stable/26225912.

Sellers, P., F. Hall, H. Margolis, B. Kelly, D. Baldocchi, G. den Hartog, J. Cihlar, M.G. Ryan, B. Goodison, P. Crill, K.J. Ranson, D. Lettenmaier, and D.E. Wickland. 1995. The Boreal Ecosystem–Atmosphere Study (BOREAS): An overview and early results from the 1994 field year. *Bulletin of the American Meteorological Society* 76(9):1549-1577. https://doi.org/10.1175/1520-0477(1995)076<1549:TBESAO>2.0.CO;2.

Sellers, P.J., F.G. Hall, R.D. Kelly, R.D. Kelly, A. Black, D. Baldocchi, J. Berry, et al. 1997. BOREAS in 1997: Experiment overview, scientific results, and future directions. *Journal of Geophysical Research: Atmospheres* 102(D24):28731-28769. https://doi.org/10.1029/97JD03300.

Serbin, S.P., A. Singh, B.E. McNeil, C.C. Kingdon, and P.A. Townsend. 2014. Spectroscopic determination of leaf morphological and biochemical traits for northern temperate and boreal tree species. *Ecological Applications* 24:1651-1669. https://doi.org/10.1890/13-2110.1.

Shi, J.J., E.P. Westeen, and D.L. Rabosky. 2018. Digitizing extant bat diversity: An open-access repository of 3D μCT-scanned skulls for research and education. *PLoS ONE* 13(9):e0203022. https://doi.org/10.1371/journal.pone.0203022.

Shi, Z., S.D. Allison, Y. He, P.A. Levine, A.M. Hoyt, J. Beem-Miller, Q. Zhu, W.R. Wieder, S. Trumbore and J.T. Randerson. 2020. The age distribution of global soil carbon inferred from radiocarbon measurements. *Nature Geoscience* 13:555559. doi:10.1038/s41561-020-0596-z.

Shrestha, R., and A.G. Boyer. 2019. Soil moisture data sets become fertile ground for applications. *Eos* 100. https://doi.org/10.1029/2019EO11432.

Singh, A., S.P. Serbin, B.E. McNeil, C.C. Kingdon, and P.A. Townsend. 2015. Imaging spectroscopy algorithms for mapping canopy foliar chemical and morphological traits and their uncertainties. *Ecological Applications* 25:2180-2197. https://doi.org/10.1890/14-2098.1.

Sonntag, D., T. Foken, H. Vömel, and O. Hellmuth. 2021. Humidity sensors. Pp. 209-241 in *Springer Handbook of Atmospheric Measurements*. Cham, Switzerland: Springer International.

Sousa, D., J.B. Fisher, F.R. Galvan, R.P. Pavlick, S. Cordell, and T.W. Giambelluca. 2021. Tree canopies reflect mycorrhizal composition. *Geophysical Research Letters* 48:e2021GL092764. https://doi.org/10.1029/2021GL092764.

Stokols, D., K.L. Hall, B.K. Taylor, and R.P. Moser. 2008. The science of team science: Overview of the field and introduction to the supplement. *American Journal of Preventive Medicine* 35(2):S77-S89. https://doi.org/10.1016/j.amepre.2008.05.002.

Tabak, M.A., M.S. Norouzzadeh, D.W. Wolfson, S.J. Sweeney, K.C. Vercauteren, N.P. Snow, J.M. Halseth, et al. 2019. Machine learning to classify animal species in camera trap images: Applications in ecology. *Methods in Ecology and Evolution* 10:585-590. https://doi.org/10.1111/2041-210X.13120.

Tape, K.D., J.A. Clark, B.M. Jones, S. Kantner, B.V. Gaglioti, G. Grosse and I. Nitze. 2022. Expanding beaver pond distribution in Arctic Alaska, 1949 to 2019. *Scientific Reports* 12(1):7123. https://doi.org/10.1038/s41598-022-09330-6.

Thompson, L.R., J.G. Sanders, D. McDonald, A. Amir, J. Ladau, K.J. Locey, R.J. Prill, et al.; Earth Microbiome Project Consortium. 2017. A communal catalogue reveals Earth's multiscale microbial diversity. *Nature* 551:457-463. https://doi.org/10.1038/nature24621.

Treat, C.C., A.-M. Virkkala, E. Burke, L. Bruhwiler, A. Chatterjee, J.B. Fisher, J. Hashemi, et al. 2024. Permafrost carbon: Progress on understanding stocks and fluxes across northern terrestrial ecosystems. *Journal of Geophysical Research: Biogeosciences* 129:e2023JG007638. https://doi.org/10.1029/2023JG007638.

Tucker, M.A., K. Böhning-Gaese, W.F. Fagan, J.M. Fryxell, B. Van Moorter, S.C. Alberts, A.H. Ali, et al. 2018. Moving in the Anthropocene: Global reductions in terrestrial mammalian movements. *Science* 359:466-469. https://doi.org/10.1126/science.aam9712.

VanderRoest, J.P., J.A. Fowler, C.C. Rhoades, H.K. Roth, C.D. Broeckling, T.S. Fegel, A.M. McKenna, E.K. Bechtold, C.M. Boot, M.J. Wilkins, and T. Borch. 2024. Fire impacts on the soil metabolome and organic matter biodegradability. *Environmental Science & Technology* 58:4167-4180. https://doi.org/10.1021/acs.est.3c09797.

Vélez, J., W. McShea, H. Shamon, P.J. Castiblanco-Camacho, M.A. Tabak, C. Chalmers, P. Fergus, and J. Fieberg. 2023. An evaluation of platforms for processing camera-trap data using artificial intelligence. *Methods in Ecology and Evolution* 14:459-477. https://doi.org/10.1111/2041-210X.14044

Walker, M., A. Mueller, K. Allen, P, Fenwick, V. Agrawal, K. Anchukaitis, and A. Hess. 2023. High resolution radiocarbon spike confirms tree ring dating with low sample depth. *Dendrochronologia* 77: 126048. https://doi.org/10.1016/j.dendro.2022.126048.

Wang, R., and J.A. Gamon. 2019. Remote sensing of terrestrial plant biodiversity. *Remote Sensing of Environment* 231:111218. https://doi.org/10.1016/j.rse.2019.111218.

Wang, R., J.A. Gamon, J. Cavender-Bares, P.A. Townsend, and A.I. Zygielbaum. 2018. The spatial sensitivity of the spectral diversity-biodiversity relationship: An experimental test in a prairie grassland. *Ecological Applications* 28:541-556. https://doi.org/10.1002/eap.1669.

Wang, Z., A. Chlus, R. Geygan, Z. Ye, T. Zheng, A. Singh, J. Couture, J. Cavender-Bares, E. Kruger, and P. Townsend. 2020. Foliar functional traits from imaging spectroscopy across biomes in the eastern North America. *New Phytologist* 228:494-511. https://doi.org/10.1111/nph.16711.

Wang, J.A., A. Baccini, M. Farina, J.T. Randerson, and M.A. Friedl. 2021. Disturbance suppresses the aboveground carbon sink in North American boreal forests. *Nature Climate Change* 11:435-441. https://doi.org/10.1038/s41558-021-01027-4.

Wang, Z., J.B. Féret, N. Liu, Z. Sun, L. Yang, S. Geng, H. Zhang, A. Chlus, E.L. Kruger, and P.A. Townsend. 2023. Generality of leaf spectroscopic models for predicting key foliar functional traits across continents: A comparison between physically-and empirically-based approaches. *Remote Sensing of Environment* 293:113614. https://doi.org/10.1016/j.rse.2023.113614.

Watts, J.D., M. Farina, J.S. Kimball, L.D. Schiferl, Z. Liu, K.A. Arndt, D. Zona, et al. 2023. Carbon uptake in Eurasian boreal forests dominates the high-latitude net ecosystem carbon budget. *Global Change Biology* 29:1870-1889. https://doi.org/10.1111/gcb.16553.

Westwood, M., N. Cavender, A. Meyer, and P. Smith. 2021. Botanic garden solutions to the plant extinction crisis. *Plants, People, Planet* 3:22-32. https://doi.org/10.1002/ppp3.10134.

White House. 2021. Executive Order 14008: Tackling the Climate Crisis at Home and Abroad. January 27, 2021. *Federal Register* 86(19):7619-7633.

Wieder, W.R., A.S. Grandy, C.M. Kallenbach, P.G. Taylor, and G.B. Bonan. 2015. Representing life in the Earth system with soil microbial functional traits in the MIMICS model. *Geoscientific Model Development* 8:1789-1808. https://doi.org/10.5194/gmd-8-1789-2015.

Wilkinson, M.D., M. Dumontier, I.J. Aalbersberg, G. Appleton, M. Axton, A. Baak, N. Blomberg, et al. 2016. The FAIR Guiding Principles for scientific data management and stewardship. *Scientific Data* 3:1-9. https://doi.org/10.1038/sdata.2016.18.

Willi, M., R.T. Pitman, A.W. Cardoso, C. Locke, A. Swanson, A. Boyer, M. Veldthuis, and L. Fortson. 2019. Identifying animal species in camera trap images using deep learning and citizen science. *Methods in Ecology and Evolution* 10:80-91. https://doi.org/10.1111/2041-210X.13099.

Williams, L.J., J. Cavender-Bares, P.A. Townsend, J.J. Couture, Z. Wang, A. Stefanski, C. Messier, and P.B. Reich. 2020. Remote spectral detection of biodiversity effects on forest biomass. *Nature Ecology & Evolution* 5:46-54. https://doi.org/10.1038/s41559-020-01329-4.

Wood, J., J.D. Ballou, T. Callicrate, J.B. Fant, M.P. Griffith, A.T. Kramer, R.C. Lacy, A. Meyer, S. Sullivan, K. Traylor-Holzer, S.K. Walsh, and K. Havens. 2020. Applying the zoo model to conservation of threatened exceptional plant species. *Conservation Biology* 34:1416-1425. https://doi.org/10.1111/cobi.13503.

Xiao, J., F. Chevallier, C. Gomez, L. Guanter, J. Hicke, A. Huete, K. Ichii, W. Ni, Y. Pang, Abdullah F. Rahman, G. Sun, W. Yuan, Li Zhang, and X. Zhang. 2019. Remote sensing of the terrestrial carbon cycle: A review of advances over 50 years. *Remote Sensing of Environment* 233:111383. https://doi.org/10.1016/j.rse.2019.111383.

Yang, Y. 2021. Emerging patterns of microbial functional traits. *Trends in Microbiology* 29:874-882. https://doi.org/10.1016/j.tim.2021.04.004.

Yoseph, E., E. Hoy, C.D. Elder, S.M. Ludwig, D.R. Thompson, and C.E. Miller. 2023. Tundra fire increases the likelihood of methane hotspot formation in the Yukon–Kuskokwim Delta, Alaska, USA. *Environmental Research Letters* 18(10):104042. https://doi.org/10.1088/1748-9326/acf50b.

Yu, D., L. Zhou, X. Liu, and G. Xu. 2023. Stable isotope-resolved metabolomics based on mass spectrometry: Methods and their applications. *TrAC Trends in Analytical Chemistry* 160:116985.

Zeng, L., B.D. Wardlow, D. Xiang, S. Hu, and D. Li. 2020. A review of vegetation phenological metrics extraction using time-series, multispectral satellite data. *Remote Sensing of Environment* 237:111511. https://doi.org/10.1016/j.rse.2019.111511.

Zeng, Y., D. Hao, A. Huete, B. Dechant, J. Berry, J.M. Chen, J. Joiner, C. Frankenberg, B. Bond-Lamberty, Y. Ryu, J. Xiao, G.R. Asrar, and M. Chen 2022. Optical vegetation indices for monitoring terrestrial ecosystems globally. *Nature Reviews Earth and Environment* 3:477-493. https://doi.org/10.1038/s43017-022-00298-5.

Zizka, A., F. Antunes Carvalho, A. Calvente, M.R. Baez-Lizarazo, A. Cabral, J.F.R. Coelho, M. Colli-Silva, et al. 2020. No one-size-fits-all solution to clean GBIF. *PeerJ* 8:e9916. https://doi.org/10.7717/peerj.9916.

5

Training and Capacity Building to Enable Continental-Scale Biology

In an age characterized by rapid technological progress and unparalleled data accessibility, the call for comprehensive training and capacity building within the scientific realm has proven clear. In turn, investing in training across critical domains is essential to arm professionals with the requisite tools and perspectives to confront the multifaceted challenges on the horizon (See Box 5-1).

In particular, three key areas of training are most relevant to developing a scientific workforce with the knowledge and skills necessary to address future challenges in CSB: **data literacy, interdisciplinary team science, and promoting diversity, equity, inclusion, and accessibility**. These are not unique to CSB, but effective development of this field is particularly dependent on them. Large-scale spatial and temporal data and data across scales are essential in CSB research; thus, data literacy, from basic to high-level expertise, will be necessary across team members to ensure an efficient workflow. Furthermore, systems thinking that emphasizes processes, relationships, feedbacks, and synthesis is critical to research advancements in CSB. Teamwork involving several disciplinary expertise and skill sets is also inherent in CSB research, such that effective communication and productive interactions across team participants with different backgrounds will be critical to ensure successful project outcomes. To maximize creativity and productivity, team science for CSB also requires inclusivity and diverse perspectives.

Across these three areas of training, the committee reviewed historic and current training efforts and discussed challenges to training a future workforce proficient in knowledge and skills related to connecting ecosystem function, resilience, vulnerability, connectivity, and/or sustainability research from small scale to regional and continental scale, and vice versa. In the context of this chapter, we define training broadly, including traditional scientific educational experiences (e.g., K-12, undergraduate, graduate, postdoctoral training), as well as less traditional educational experiences (e.g., post-baccalaureate experiences, public outreach, short-term internships, team training). Distinct from but complementary to training is the need for capacity building, which we define as increasing the ability and resources of the overall scientific endeavor to effectively and sustainably support research activities. The committee aimed to evaluate and create recommendations on training and capacity building for CSB for the research community, funders, and decision makers. To be able to inform responses to global environmental crises, the workforce must be able to perform research at multiple scales in the most effective way. The committee determined that training exists in these three core areas stated above, but they are rarely woven together in a way that would most robustly support CSB (Figure 5-1).

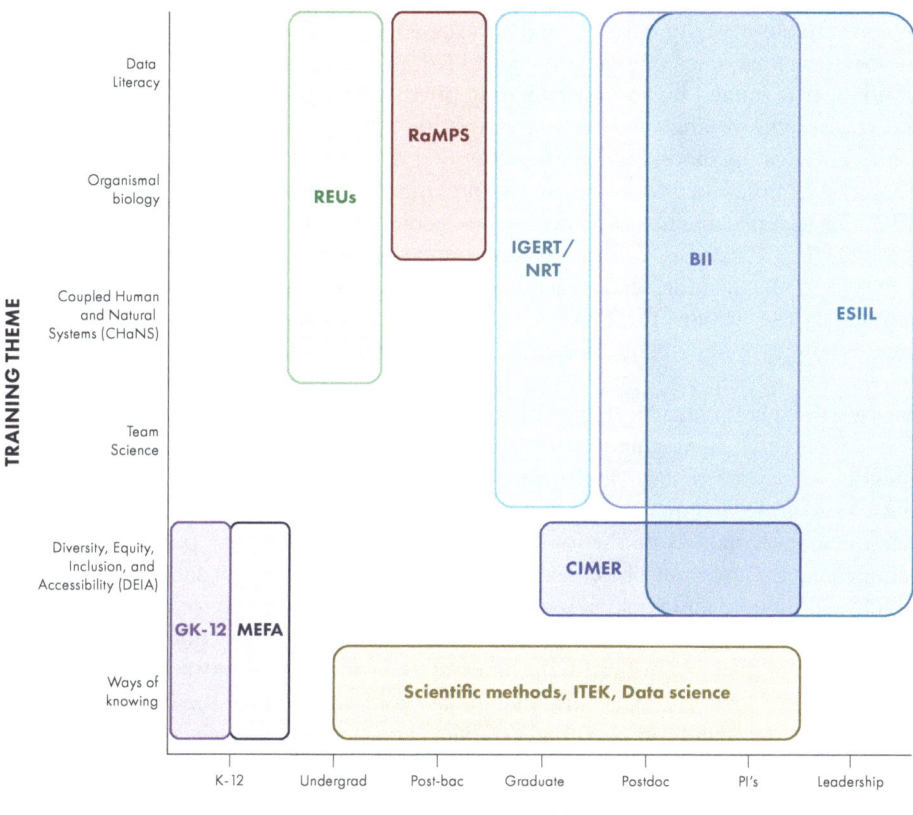

FIGURE 5-1 Numerous current and former training programs exist to support a scientific workforce. They can provide the skills and knowledge training necessary to address CSB research, but gaps, highlighted by areas of white space, exist.
NOTE: Programs are examples and do not comprehensively encapsulate all programs that can address CSB research. BII: NSF Biology Integration Institutes, CIMER: Center for the Improvement of Mentored Experiences in Research, ESIIL: Environmental Data Science Innovation & Inclusion Lab, IGERT: NSF Integrative Graduate Education and Research Traineeship, NRT: NSF Research Traineeship, ITEK: Indigenous Traditional Ecological Knowledge; RaMPS: NSF Research and Mentoring for Postbaccalaureates in Biological Sciences, REUs: Research Experiences for Undergraduates.

DATA LITERACY TRAINING AND CAPACITY BUILDING

In a practical sense, addressing biological questions at continental scales requires data-intensive and team science approaches (Cheruvelil and Soranno. 2018). Relevant data generated in the laboratory, field, from remote sensors, or through modeling are diverse and associated with domain-specific norms in terms of format and archive. The data skills needed are not specialized to CSB, and these skills have been called for across various domains in the past (Carroll et al. 2021). In particular, there is a need for developing an understanding of statistics at the scale of big data. Proficiency in these skills is paramount in this context due to the functional necessity of a data-savvy workforce in CSB. However, not all individuals need to be experts in all areas of data handling and analysis. Rather, CSB researchers should possess basic competencies and be comfortable working in teams and accessing each others' expertise.

Six principal areas of skill are needed for a data-literate workforce: (1) data collection that is framed by an understanding of data management; (2) data management and processing, (3) analysis, (4) software skills for science, (5) visualization, and (6) communication methods for collaboration and dissemination (Hampton et al. 2017). Chapter 4 explores how formal networks provide many of these skills, specifically data collection and management and methods for collaboration. Individual researchers can combine their expertise in these different areas through collaborating within formal networks to alleviate the need to acquire expertise in all these areas. At the same time, individuals can acquire these skills and develop connections with other skilled colleagues so they will be better poised to develop future networks of their own.

Collecting data with an understanding of data management means considering long-term needs for standardizing data and metadata at the point of collection, as well as excellent method documentation for reproducibility, reusability, and meaningful integration and inference. Data management requires understanding basic data formats, versioning, and quality assessment, as well as standards for metadata that enable data integration and ensure the long-term value of the data. At the point of collection, data generators also could be considering long-term data storage, access, and sharing, which will involve an investigation of the appropriate data repositories. For example, one expects to find most forms of genetic data in GenBank,[1] while many forms of environmental data are in the Environmental Data Initiative.[2] Storing data in repositories with metadata and data formats that are gold standard for the domain will help to maximize the long-term usefulness of data. The standard for data generators can be that they make their data Findable, Accessible, Interoperable, and Reusable (FAIR). Relevant analyses will depend on the scientific questions, and the potential analyses that can be employed are continually expanding—for example, through Bayesian and machine learning approaches—such that a full review of statistical analyses here would be quickly outdated. Rather, the committee suggests focusing on a set of skills that underpin a variety of analyses: simulation, sampling, visualization, and summary statistics. Scripting in any computational language (e.g., R, Python) captures the sci-

---

[1] See https://www.ncbi.nlm.nih.gov/genbank/ (accessed February 7, 2024).
[2] See https://edirepository.org/ (accessed February 7, 2024).

> **BOX 5-1**
> **Connecting Core Themes to Training and Capacity-Building Efforts Used in Research Across Scales**
>
> In addressing training capabilities, we are drawn back to the core themes that underpin CSB: Connectivity; Resilience and Vulnerability; Biodiversity and Ecosystem Function; and Sustainability of Ecosystem Services. These themes not only guide research endeavors but also underscore the fundamental competencies required—data literacy, interdisciplinary collaboration, and diversity, equity, and inclusion. This chapter serves as a conduit for aligning training efforts with these themes, emphasizing the importance of developing a workforce equipped to understand the intricate interconnections within and between ecosystems (Connectivity), navigate and respond to environmental disturbances (Resilience and Vulnerability), comprehend the deep understanding of ecosystems functioning (Biodiversity and Ecosystem Function), and ensure systems thinking grounded around sustainability initiatives (Sustainability of Ecosystem Services). By fostering competencies in data literacy, interdisciplinary collaboration, and diversity, equity, and inclusion, the groundwork is laid for addressing these themes effectively, enabling researchers to navigate the complexities of ecosystem dynamics and devise holistic, sustainable solutions to global environmental challenges.

entific workflow from data ingestion to data analysis and visualization. Many, but not all, subdomains of biology routinely teach undergraduates to program, and the standard languages differ across fields. The knowledge and agility one gains in one language can help them to operate in other languages that are less familiar (Videnovik et al. 2010). As analyses are scaled up, biologists also need to become more familiar with best practices in software development (e.g., versioning; Wilson et al. 2014), and become comfortable with online resources through which they can benefit from others helping create more effective code (e.g., Github). For example, analyses that scale up from a desktop computer to high-performance computing may benefit from parallelization, and large-scale data integrations or complex analyses may gain efficiencies from more formalized scientific workflows (Farley et al. 2018). Data visualization aids all stages of CSB, from checking the quality of data and analyses, to the final communication of results. Finally, it is clear that communication skills need to be gained alongside technical skills in working with data in order to be maximally effective. No one needs to be an expert in all the skills that underpin CSB; rather, individuals need to be comfortable working in a team, and prepared with the skills for effective communication that make teamwork successful (Cheruvelil and Soranno 2018, Cheruvelil et al. 2014, NRC 2015).

Enhancing data literacy means not only elevating the skills of those researchers working at the leading edges of CSB, for example, those funded by the NSF initiatives listed in Box 1-1, but also those currently lacking basic data skills of any kind, working

> **BOX 5-2**
> **Examples of Training and Capacity-Building Efforts Used to Successfully Connect Research Across Scales**
>
> In the pursuit of advancing scientific research, the integration of data across various scales is paramount. To achieve this goal, numerous training and capacity-building efforts have been implemented, spanning from workshops and online resources to specialized graduate programs. These endeavors aim to equip researchers with the necessary skills and tools to navigate the complexities of contemporary scientific inquiry. The following examples highlight various initiatives that have successfully fostered collaboration and innovation in connecting research across scales.
>
> *Workshops:*
> - Workshops, such as "Train the Trainer" sessions offered by Software Carpentry and Data Carpentry, have proven invaluable in enhancing data management and analytical skills.
> - Participation in the Near-term Ecological Forecasting Summer Institute has provided researchers with practical forecasting techniques and tools.
> - The National Center for Ecological Analysis and Synthesis CoreR workshops have equipped participants with advanced skills in ecological data analysis and synthesis.
>
> *Online resources:*
> - Macrosystems EDDIE enables researchers to access cutting-edge tools and methodologies for analyzing large-scale ecological data.
> - University of Colorado Boulder Earth Lab provides free online resources for environmental informatics, complemented by an optional certification program, fostering proficiency in data analysis techniques.

throughout the domains upon which CSB is built (some have referred to this principle as a "rising tide lifts all boats" or "raising the floor"). Basic data literacy is increasingly important across all domains and sectors, and in individuals' lives as well, such that these skills are transferable both to the workplaces and to making informed decisions as citizens (UNESCO 2006). Some have argued that data literacy should be incorporated across the curriculum (e.g., Kjelvik and Schultheis 2019), rather than simply as stand-alone workshops and courses, in order for students to appreciate the authentic experiences of using data in the context of real-world questions and situations (Kjelvik and Schultheis 2019, Langen et al. 2014). "Data across the curriculum" could be modeled after "writing across the curriculum" widely implemented in universities since proposed in the 1970s; the idea is that writing is not only a useful skill set but also helps one think (Hampton et al. 2017). The same argument could be made for data skills (See Box 5-2).

In many ways, the focus on data necessitates new ways of thinking, and presents some unresolved questions about how science is done. First, scientists have been unac-

- Cross-sectional data and ecological university courses, such as Emilio Bruna's data collection and management courses,[a] and Ethan White's data carpentry for biologists.[b]

Graduate programs integrating informatics and ecology:
- Northern Arizona University's Ecological and Environmental Informatics doctoral program integrates advanced informatics techniques with ecological research methodologies, preparing graduates for interdisciplinary research.
- The University of California National Center for Ecological Analysis and Synthesis Master of Environmental Data Science program equips students with a comprehensive understanding of environmental data analysis and management, preparing them for careers at the intersection of ecology and informatics.

Specialized programs:
- The Data Science for Energy and Environmental Research program at the University of Chicago offers specialized training in data science techniques tailored for energy and environmental research applications.
- University of California Berkeley's Environment and Society: Data Sciences for the 21st Century program provides students with the skills necessary to address complex environmental challenges using data-driven approaches.
- Northwestern University's Integrated Data-Driven Discovery in Earth and Astrophysical Sciences (IDEAs) program fosters interdisciplinary collaboration and innovation in the analysis of Earth and astrophysical data.

[a] See https://tropicos.netlify.app/courses/las6292-data/ (accessed April 28, 2024).
[b] See https://datacarpentry.org/semester-biology/ (accessed April 28, 2024).

customed to thinking of data as a scientific product. Rather, it has been viewed as a "precursor to publication" (Elliot et al. 2016; Hampton et al. 2013). These attitudes are changing, evidenced by recent changes in funders' policies, such as requiring data management plans and data sharing. This relatively new awareness ideally will spawn more robust data management approaches, cultivate wider data literacy, and the broader availability of and CARE (Collective benefit, Authority, Responsibility, Ethics) data that can be integrated in large-scale analyses. Perhaps more difficult to describe is the shift in mindset occurring about how "big data" are used in the scientific process, in which the gold standard is the testing of ideas based on a priori hypotheses (Sterner and Elliot 2022). Students are frequently confused about how and when to test a hypothesis with the data that are available (e.g., Langen et al. 2014). Are they allowed to look at the data to see if it is appropriate for their hypothesis testing? Is it wrong to change their hypotheses if they see that the dataset is not appropriate for testing? How much are they allowed to interrogate the data before they are "guilty" of "fishing" or "p-hacking"?

Many scientists engaged in data-intensive analyses see big datasets as new lenses on the world. It has long been accepted that we can develop hypotheses about the distribution of trees or animals based on our visual observations of nature, before designing experiments and collecting quantitative data. If data provide the primary lens through which one can "see" the patterns of the wind or microbes, otherwise invisible, it is hard to argue against using such data to inspire new hypotheses. Perhaps a key principle is that such investigations are guided by sound theory, as discussed in Chapter 3.

Challenges That Limit Training and Capacity-Building Efforts to Connect Research Across Scales

In the pursuit of interdisciplinary collaboration and innovation, addressing the challenges that hinder training and capacity-building efforts is increasingly important. One such challenge is the saturation of existing curricula, which limits the incorporation of additional training modules focused on essential data skills necessary for connecting research across scales. Furthermore, instructors in the environmental sciences often face challenges in integrating data skills into their teaching, potentially inhibiting the dissemination of crucial knowledge. Moreover, the rapid pace of technological advancement outpaces researchers' capacity to adapt and acquire the requisite data analysis skills, creating a gap between technological innovation and skill acquisition. These challenges underscore the need for targeted interventions and support mechanisms to overcome barriers and foster effective training and capacity-building initiatives in the scientific community.

Conclusions on Training and Capacity-Building Efforts to Connect Research Across Scales

To confront the challenges inherent in CSB, the committee presents conclusions surrounding training and capacity-building efforts aimed at equipping and connecting research across various scales. Emphasizing the need for refined training methodologies, the focus lies on elevating the efficacy of biological research endeavors. At the heart of this pursuit lies the cultivation of a collaborative research environment, where team members possess the agility to traverse diverse disciplines and harness multifaceted skill sets. As CSB ventures continue to transcend disciplinary boundaries, fostering a cohort of intersectional researchers capable of navigating these complexities emerges as imperative. Within this exploration, key areas of data literacy training are identified below, underscoring the pivotal role of interdisciplinary collaboration and adaptability in advancing biological inquiry.

Conclusion 5-1: Training aimed at individuals addressing CSB research. Individuals can actively seek opportunities for upskilling, such as workshops and online resources, and advocate for engagement in upskilling among students and colleagues. Additionally, they could integrate data projects into teaching, following examples such as Macrosystems EDDIE initiatives.

Conclusion 5-2: Training across curricula of institutions. Institutions can incorporate courses covering essential data science aspects, including data collection with a focus on efficient data management, data processing, advanced analysis techniques, software skills tailored for scientific applications, data visualization, and effective communication methods for collaborative projects and dissemination. They can also promote faculty upskilling in data management and programming and establish mechanisms to recognize and incentivize the publication of data as a legitimate research output. Analogous to the way many universities aim to develop writing skills through multiple courses and majors employing a "writing across the curriculum" approach, institutions could explore the adoption of a "data across the curriculum" approach to coordinate data science knowledge and skills across a variety of courses and degree programs.

Conclusion 5-3: Training aimed at strengthening federal collaboration. Agencies and institutions could address the current need for resources allocated to training a proficient data-savvy workforce compared to the escalating demand. In accordance with the 5-year federal STEM Education Strategic Plan for 2018–2022 (and the 2023-2028 plan under development), they could provide robust support for education and training initiatives focusing on environmental data skills and fostering continental-scale thinking, exemplified by programs such as Macrosystems education and training. Additionally, they can consider allocating targeted resources to bolster research endeavors utilizing data from observatories such as the National Ecological Observatory Network (NEON), fostering the integration of data skills into research endeavors.

INTERDISCIPLINARY TEAM SCIENCE TRAINING

Interdisciplinary teams are an essential component of working across biological scales where collaborative research is fundamental to integrating knowledge and resources (NRC 2004). This interdisciplinary approach is driven by the recognition that complex research questions often require input and expertise from various domains and skill sets, for example, organismal biology, ecosystem ecology, biogeography, remote sensing, and data sciences (NASEM 2023). Overall, interdisciplinary team projects are more productive and innovative than individualistic disciplinary ones (Hall et al. 2018), and collaborative teams of scientists have become essential in tackling multifaceted challenges and driving innovation (Fiore 2008, Fortunato et al. 2018; NRC 2015). Team science, defined as research conducted by more than one person in an interdependent fashion, plays a pivotal role in developing protocols that ensure effective collaborative work. Most teams currently tackling macrosystems problems that will be at the heart of CSB are formed by researchers from different disciplines, frequently from different geographic areas and institutions, and involve participants at several career stages (Dodds et al. 2021, Read et al. 2016); as such, team science training becomes critical to ensure a project's success (Cheruvelil et al. 2014).

Team science training, that is, training in the disciplinary, communication, and interaction skills needed to work as part of an interdisciplinary team, are critical for these teams. Within these collaborative teams, each member not only contributes his or her specialized knowledge but must also be proficient in exchanging information across disciplinary boundaries. Effective teamwork in interdisciplinary projects demands a certain level of literacy in other fields and skill sets to facilitate connections and bridge the gaps between different areas of expertise. Even before the work starts, research networking tools can be used not only to assess any gaps among initial participants but also to foster connections and identify new ones among them (Vacca et al. 2015).

Team participation will likely span from undergraduate students embarking on their first research projects to senior researchers with extensive experience. Therefore, providing team training at all career stages is vital, with a particular emphasis on early-stage researchers who are just entering the collaborative research landscape (Read et al. 2016). Additionally, navigating the dynamics of a diverse team and having the ability to negotiate and resolve conflicts are crucial skills for all team members. To maximize team function in CSB—that is, defining roles and responsibilities, setting expectations, and developing policies for authorship and data sharing—team science training will be critical (Cheruvelil and Soranno 2018). Many of these traits are outlined in Chapter 4 as characteristics of successful networks, notably within the context of small-scale team science successfully growing into formal networks. Regardless of how a network forms, collaboration is key to successful network product formation, including data among many others.

This emphasis on team science training is echoed in a report from the National Academies of Sciences, Engineering, and Medicine (NRC 2015), which highlights the importance of cultivating collaboration and interdisciplinary skills in the scientific community. In the ecological field, researchers have also emphasized the need for team science training (e.g., Peterson et al. 2023, Read et al. 2016), underlining its relevance across scientific disciplines. This training could focus on professional competencies that ensure that team members possess a certain level of knowledge of the biological system, skills necessary to process and analyze associated biological and ecological data, and attributes necessary to make teamwork productive (Wiek et al. 2015). Training topics and methods can be geared to the team's specific needs; however, team generic competencies include:

1. **Basic literacy:** training team members on the basic competencies—that is, disciplinary knowledge and skills—is necessary for optimal interactions. Such training will enable team members to understand the contributions of other team members (Stagl et al. 2007). Cross-training will improve each individual's understanding of the team's objectives (Fiore et al. 2005) and their contribution to those objectives. In this context that would include basic knowledge of:
   a. *Disciplinary knowledge of the systems under study as well as data collection design and protocols.* A basic understanding of the systems in question is critical for developing the analytical tools that best represent those systems. Exchange of interdisciplinary/interskills knowledge is commonly accomplished by close interactions between system and analytical experts, and a certain level of literacy in the other's discipline will facilitate communication and help advance the project's objectives.

b. *Data sciences.* The rise of data-intensive science in biological studies across scales will require team members to have basic methodological knowledge of data processing, analysis, and visualization. Basic training in data sciences—including statistics, data management, data engineering, and computer languages—will be required not only to manage and use complex data (see section on Data Literacy Training and Capacity Building above), but also to improve team members' ability to recognize linkages and assess possibilities.

c. *Modeling approaches.* As with data sciences, basic knowledge of the models being used and developed will improve the capacity of team members to investigate possibilities, identify knowledge gaps, and design scientific advances (Pennington et al. 2020). Visual representations of the analytical methods may not only aid with development and interpretation, but it may also facilitate innovation by involving different conceptualizations and varying perspectives of the problem, experiment, and/or hypothesis at hand (Nersessian 1999).

Most of this learning can take place during the collaboration (Pennington et al. 2013) and include developing the necessary vocabulary to communicate across disciplines and skill sets (Pennington et al. 2020).

2. **Communication:** Effective communication skills are vital for conveying ideas and information across disciplines and skill sets. Starting with creating a shared vision, information acquisition, collaboration, and dissemination all play an important role in team performance (Xia and Ya 2012). Language training and team management skills help team members transfer knowledge to colleagues outside their areas of expertise, facilitating productive collaboration (IOM 2005). Being able to draw, evaluate, modify, and integrate insights are skills that team members should have (Newell and Luckie 2019). As an example, the Employing Model-Based Reasoning in Socio-Environmental Synthesis (EMBeRS; Pennington 2016, Pennington et al. 2021) was developed to design and test collaborative knowledge integration via model-based reasoning. It includes modules on team composition and team processes and addresses disciplinary differences.

3. **Team dynamics:** Cooperative work requires a certain degree of "social intelligence" due to the complex social and intellectual processes involved. Training in team dynamics helps team members understand and navigate the intricacies of working together effectively (Fiore 2008). Furthermore, coordination of team tasks and frequent communication are critical in the production of new knowledge and tools and successful training of students (Cummings and Kiesler 2007). Leadership and project management skills training, across career stages, could also benefit the dynamics and productivity of the team. Mentorship training will also benefit team dynamics, and programs such as the Center for the Improvement of Mentored Experiences in Research (CIMER[3]) can provide resources for improving mentoring relationships.

---

[3] See https://cimerproject.org/ (accessed April 28, 2024).

4. *Negotiation and conflict resolution:* Conflict is a natural part of collaborative work, and the ability to negotiate and resolve conflicts constructively is essential for maintaining a harmonious and productive team environment. Trust among team members is crucial for having open discussions about the project that do not lead to conflict or misinterpretation (Bennett and Gadlin 2012). Thus, specific steps to establishing trust—e.g., conversations about tasks assignments, authorship, data sharing, and decision making—can be one of the first tasks during team training (Bennett and Gadlin 2012). Ensuring that all team members are well versed in conflict management strategies may be the key to a successful progression of work. Conflict among team members can be particularly damaging when there are career-associated power differences, for example, student–advisor (Brockman et al. 2010). Interest-based and cooperative approaches may work best in settings with a diversity of career stages (Klomparens et al. 2008).

In addition to these fundamental aspects of team science training, there are other areas of training that are intrinsically connected to research in CSB:

1. *Connection to theory:* Understanding the broader context of biological and ecological research, including the role of theory and the importance of advancing science, provides purpose and direction to interdisciplinary projects (see Chapter 3). Being able to assess what level of theoretical simplicity is sufficient to assess how scales relate will be key to developing and advancing CSB science.

To promote discovery and better address environmental challenges, data analyses should be guided by specific goals developed in accordance with current understanding of the system and knowledge needs. For that, substantial theoretical knowledge by team members that work closely with other members will be critical. Forecasting trends and predicting thresholds will require CSB research to complement pattern recognition with mechanistic processes and synergies underlying those patterns. Training team members to move beyond data exploration and to be able to formulate knowledge-based research goals will be essential for the development of CSB science. Given the complexity of CSB science, linking theory and going back and forth between theory and empirical work are vital. Particularly, early-career researchers will greatly benefit from learning to develop theoretical frameworks, relate frameworks, test frameworks with data, and communicate conclusions to other CSB members. To effectively assess ecosystem functioning—its resilience and vulnerability, and ultimately the sustainability of the ecosystem services provided—will require a conceptualization that links processes and drivers acting across scales.

2. *Systems thinking:* The nature of CSB makes training on systems thinking a potential asset for its success. Understanding the many biological processes involved, their differences and relationships, connections and feedbacks, their synthesis as a whole, are all critical concepts in CSB sciences. Training

researchers from a modular mindset to high-conductivity thinking will also benefit cross-scale research (Elmqvist et al. 2021). Systems thinking training provides tools that facilitate this integrated conceptualization of a system. Furthermore, in many research contexts, collaborating with stakeholders or partners from industries, agencies, or local governments and communities is essential. Educational programs that integrate such collaborations—such as those within NEON and Google Earth, for example—can be valuable resources for interdisciplinary research teams.

Interdisciplinary work and team science are essential for addressing complex research challenges associated with working at multiple scales. Effective training in communication, literacy, team dynamics, negotiation, and conflict resolution is crucial for the success of collaborative research projects. Additionally, training to use and develop theoretical knowledge will ensure that research is led by the goal of advancing biology, and systems-thinking training that incorporates the intricacies of CSB in engaging with coupled human–natural systems can further enrich interdisciplinary teams and contribute to the advancement of science.

CHALLENGES THAT LIMIT EFFECTIVE RESEARCH ACROSS SCALES AND TRAINING APPROACHES TO OVERCOME THEM

Identifying and addressing challenges in CSB associated with training will be relevant to improving research efforts. The committee identified three major challenges associated with team science training. First, effective collaboration practices are crucial for advancing research in CSB. Encouraging collaborative efforts and breaking down silos are essential to foster teamwork and ensure a collective approach toward research objectives. Second, communication across three dimensions is imperative in CSB endeavors. This includes improving communication across disciplines, skill levels, and career stages to facilitate the productive exchange of ideas and expertise. Last, coordinating disjointed teams scattered across different institutions and geographical locations presents a significant challenge. Overcoming these obstacles requires establishing efficient coordination mechanisms to bridge gaps and streamline efforts for more effective research outcomes.

Conclusions on Interdisciplinary Team Science Training to Connect Research Across Scales

Creating a more adaptable and collaborative research landscape where team members can navigate between different disciplines and skill sets will be essential to the success of CSB. Areas of team science training that will enhance the overall effectiveness of CBS research projects are:

Conclusion 5-4: Training aimed at developing comprehensive theoretical frameworks addressing CSB research: Funding agencies and institutions can provide research networking tools that encourage cross-expertise sharing and communication among teams, exemplified by initiatives such as Systems Thinking training, the Cornell Systems Thinking Certificate, and programs such as EMBeRS.

Conclusion 5-5: Training aimed at enhancing understanding of team roles. Funding agencies and institutions could provide guidelines and resources for cross-training to enhance comprehension of each team members' role, incorporating initiatives such as interdisciplinary language training and advanced communication skill development.

Conclusion 5-6: Training aimed at strengthening collaboration. Funding agencies and institutions can actively promote and furnish guidelines and resources for comprehensive team science training. This may encompass diverse training modules such as negotiations and conflict management, leadership skills, project management, and fostering partnerships with collaborators.

Conclusion 5-7: Training aimed at supporting across career-stage training. CSB teams may provide comprehensive guidelines and resources for training across various career stages. This includes offerings such as employment training, participation in mentoring programs such as CIMER, and involvement in prestigious programs such as the Earth Leadership Program, GLEON Fellowship program, and NSF Research Traineeships.

Conclusion 5-8: Training aimed at ensuring effective communication in disjointed teams: CSB teams can establish guidelines and resources aimed at facilitating effective communication and ensuring workflow continuity among participants operating across different institutions and/or geographic areas. For instance, implementing strategies to ensure seamless communication and collaboration despite physical or organizational distances is important.

EVIDENCE-BASED METHODS TO PROMOTE DIVERSITY, EQUITY, INCLUSION, AND ACCESSIBILITY IN CONTINENTAL-SCALE BIOLOGY

Across academia (Stewart 2021, Zhu et al. 2021), government (Hofstra et al. 2020, Nielsen et al. 2017, White House 2021a), industry (AECOM 2022), and nongovernmental organizations (National Council of Nonprofits 2022), there is an agreed-upon conclusion about the need to diversify the science, technology, engineering, mathematics, and medicine (STEMM) workforce, and to create more inclusive and equitable workplace environments to retain this workforce (NASEM 2021).

NASEM (2023) defines diversity as the fair representation of different human characteristics and perspectives within a group, emphasizing the contextual nature of diversity and its importance in specific contexts. Equity is described as the outcome of fair conditions that provide all individuals and groups with the resources needed for general well-being or success, distinguishing it from the concept of equality. Inclusion refers to the sense of belonging in an environment where individuals feel supported and have a voice. This framework is vital for CSB because it ensures that diverse perspectives and backgrounds are incorporated, fostering a richer understanding of biological systems across vast geographical areas.

Additionally, accessibility plays a critical role, focusing on the design and provision of facilities and information to enable all individuals, including those with disabilities, to fully participate (NASEM 2024). Despite legal protections, such as the Americans with Disabilities Act, which guarantee access to education and employment in STEMM fields, people with disabilities remain underrepresented. The 2023 NASEM report emphasizes the need for antiracism and comprehensive diversity, equity, and inclusion (DEI) initiatives in STEMM to dismantle systemic barriers and promote equitable opportunities, outlining specific recommendations for policy changes, institutional practices, and leadership strategies to foster inclusivity and accessibility in scientific and educational environments. The emphasis on accessibility, which is expressed in further detail below, is particularly critical in the context of CSB, ensuring that individuals with diverse abilities have equal opportunities to contribute to and benefit from the understanding of vast ecosystems and biodiversity across continents.

Previous NASEM consensus studies have examined different components of increasing inclusivity in STEMM (e.g., women of color in tech, research at minority-serving institutions), but the 2023 *Beyond Broadening Participation* report is the first and most comprehensive consensus study to examine antiracism and DEI holistically across STEMM. NASEM (2023) reviewed bias and racism in STEMM workplaces, proposed strategies to enhance DEI, and emphasized antiracism as active measures against systemic racism. The report's recommendations include requests for increased support for minority-serving institutions, evidence-based programs to connect minoritized individuals, and leadership responsibilities for advancing DEI.

DEI Themes Particularly Relevant to Continental-Scale Biology

In addition to the general report outlined above, the committee noted that the interdisciplinary nature of CSB necessitates inclusivity, diversity, and evidence-based best practices in team science. At broad spatial and temporal scales, humans are undeniably a major driver of pattern and process, and inequities in socioeconomic systems are reflected in natural systems. Executive Order 14096 encourages federal activities across the whole of government—including those related to science, data, and research—to advance environmental justice, the just treatment and equal involvement of everyone, regardless of income, race, color, national origin, Tribal affiliation, or disability—regarding environmental protections and benefits, as well as meaningful involvement in the policies that shape their communities. OSTP recently released the first Environmental Justice Science, Data, and Research Plan which charges the scientific community with providing critical evidence that federal agencies can use to develop environmental justice policies and decisions, ensuring that actions are informed, targeted, and effective (National Science and Technology Council 2024). Research and training in CSB are central to achieving the science, data, and research goals that play important roles in the achievement of environmental justice. Indeed, the CSB research community is increasingly calling for incorporating environmental justice into telecoupling research to better address issues surrounding socioecological inequities with common governance across distances (Boillat et al. 2020).

Given CSB's interdisciplinary nature, embracing inclusivity and diverse perspectives is essential to foster innovation and productivity. However, interdisciplinary research is not always valued by traditional academic systems of career advancement (e.g., promotion and tenure). It can be risky for pretenure faculty to engage in research that addresses socioecological systems if they were hired to conduct primarily research (and vice versa) or to attempt work at larger spatial or temporal scales. This risk is particularly high for faculty at moderate research activity (i.e., R3) or primarily undergraduate institutions who are more often "the only" in their discipline or for faculty from minoritized or underrepresented groups. Funding programs that focus on principal investigators at institutions such as the Building Research Capacity of New Faculty in Biology (BRC-BIO) program support pretenure faculty by demonstrating to their universities that they are engaging in cutting-edge, funding-worthy research. Furthermore, meetings that bring CSB researchers together and elevate the work of early-career scientists (e.g., ESIIL Innovation Summit) help grow CSB research networks and build critical mass among researchers, which lends validity to the work by tenure and promotion committees and tenure letter writers. As much as possible, these meetings could be open and accessible to all current and interested CSB research community stakeholders.

Three additional themes, described as (1) Accessibility, (2) Indigenous and Traditional Ecological Knowledge, and (3) Systemic and Cultural Change, were found to be particularly relevant to CSB and are described below:

Accessibility

Accessibility has emerged as a significant concern highlighted by experts in DEI within STEMM. These concerns pertained to access for both individuals with disabilities and people belonging to other historically minoritized groups in STEM. The interdisciplinary nature of CSB means that students and trainees enter CSB through a variety of pathways given their varied disciplines. Internships, research assistantships, or other traineeships may be in field experiences, data science, museum collections, socioeconomic research, and more. Field experiences (e.g., seasonal internships with NEON) may not be accessible to individuals with disabilities or for whom identity poses safety concerns in certain locations due to race/ethnicity, sexual orientation, gender identity, and/or religion (Demery and Pipkin 2021). On the data science side, experiences are not always accessible to people with certain disabilities because websites and data standards (i.e., computer languages, data visualization software, etc.) vary in their accessibility and adherence to universal design. Solutions exist that make training and employment opportunities across CSB more accessible for individuals from a variety of marginalized backgrounds. For example, clear community guidelines that utilize gender-inclusive language, established medical/emergency response systems, and developed plans for travel and accommodation create an inclusive environment for members of the LGBTQ+ community and demonstrate a commitment from project leadership (Lundin and Bombaci 2022). For individuals with disabilities, accommodations such as flexible schedules, remote work options, 508 compliance, and a supportive work culture support access to CSB training and employment. During the Covid-19 pandemic, many individuals benefited from and relied on technological advances championed by disability advocates; continuing to accommodate people with disabilities in STEM benefits us all (Daehn and Croxson 2021).

In one of the committee's public information-gathering sessions on inclusive training and workforce development, expert panelist Dr. Sara Bombaci also identified limited income opportunities as a significant financial barrier to diversifying and retaining a diverse workforce for CSB. In a nationwide survey of undergraduate students interested in environmental and natural sciences jobs, 43 percent of respondents agreed or strongly agreed that income was a barrier to accepting an internship (Jensen et al. 2021). Students reported that, with additional support for housing and transportation, a position would need to pay $8.68/hour in order for them to accept it, but only 65 percent of jobs (according to job board surveys) pay $8.68 or more. Students who identified as racial and/or ethnic minorities reported that they needed $10.80/hour but only 56 percent of jobs paid $10.80/hour or more. This could be because many racially or ethnically minoritized students lack a financial "safety net" provided by family members or because they themselves financially support family members. An hourly rate of $20.00/hour was required to retain 90 percent of all students, but only 3 percent of jobs in the environmental and natural sciences, including CSB, paid $20.00/hour or more. Researchers also identified other key barriers to recruiting and retaining a diverse workforce including conflicts with work and school, lack of transportation and housing, and mental health or physical concerns about being able to carry out the work (Jensen et al. 2021). Finally, the seasonal nature of certain positions emphasized in CSB is a barrier for applicants who do not want or cannot afford short-term work.

Indigenous and Traditional Ecological Knowledge

At the outset of its formation, the committee recognized the key role that Indigenous and Traditional ecological knowledge (ITEK) plays in CSB. ITEK is a body of observations, oral and written knowledge, practices, and beliefs that promotes environmental sustainability and the responsible stewardship of natural resources through relationships between humans and environmental systems. It is applied to phenomena across biological, physical, cultural, and spiritual systems. As conceptualized, CSB addresses questions about biological processes and patterns that emerge at broad organizational, spatial, and/or temporal scales, which is inherently tied to land and its historic legacy of use and contemporary management by Indigenous peoples. As Dr. Gillian Bowser stated in her presentation to the committee, "Data is cultural, economic, and place-based." Today, Indigenous peoples remain stewards of a considerable amount of land and biodiversity on Earth; globally, Indigenous lands intersect with 40 percent of protected areas and overall cover roughly one-fourth of Earth's surface (Grantham 2022). The White House Office of Science and Technology Policy (OSTP) released a 2021 memorandum committing to elevating ITEK in federal scientific and policy processes in the United States (White House 2021b). Currently an ad hoc National Academies Committee on the

Co-Production of Environmental Knowledge, Methods, and Approaches[4] is writing a report on the nature and sociocultural dimensions of bridging and integrating Indigenous and local knowledge systems with professional scientific ones.

Systemic and Cultural Change

During the committee's third information-gathering session, 30-year STEMM equity expert and scholar Dr. John Matsui stated that programs devoted to promoting DEI in STEMM produce significant knowledge around identifying evidence-based, inclusive, and cost-effective best practices in closing STEMM equity gaps. He goes on, however, to state that programs themselves are not the solution; programs address symptoms of a broken system (e.g., equity gaps, underrepresentation) and direct services to students are "institutional workarounds." Instead, programs should be viewed as "labs" or "incubators" to produce knowledge around ways to fix institutions, not the students, and to develop talent instead of skim talent. Successful models that address systemic and cultural change, or "Inclusive Excellence," in STEMM include NSF ADVANCE (Organizational Change for Gender Equity in STEM Academic Professions), National Institutes of Health BUILD (BUilding Infrastructure Leading to Diversity), and Howard Hughes Medical Institute (HHMI) Driving Change. HHMI Driving Change is a multi-year funding initiative designed to encourage a comprehensive approach to institutional culture change using three elements: creation of a multi-institutional learning community, an institution-centered program designed to promote inclusivity in a university's STEM learning environment, and student-centered programs where faculty assume responsibility for the success of all students. Phase I of the program is a deep self-study of the university's systems and culture that contribute to equity gaps and underrepresentation. Only after a thorough self-study that demonstrates knowledge of a university's own unique challenges are universities eligible for Phase II funding that supports program implementation. The HHMI Driving Change program embodies what Dr. Matsui and many other STEMM equity experts suggest: that top-down programs are helpful, but true systemic and cultural change is local and happens at the institutional level. This is a challenge because the DEI in STEMM landscape in the United States is becoming increasingly patchy due to varied investment at state and institutional levels, funding agencies have a significant opportunity and play a key role in incentivizing systemic and cultural change and supporting the necessary staffing to do this work.

Another gap exists around training of faculty and leadership to learn equitable advising, mentoring, and teaching practices. Researchers in CSB who supervise others (e.g., faculty, team leaders, program directors) need regular training in evidence-based best practices. For example, CIMER at the University of Wisconsin–Madison aims to improve research mentoring relationships for mentees and mentors at all career stages through the development, implementation, and study of evidence-based and culturally responsive interventions. CIMER offers numerous curricula, provides training, and has

---

[4] See https://www.nationalacademies.org/our-work/co-production-of-environmental-knowledge-methods-and-approaches (accessed February 7, 2024).

trained almost 1,000 facilitators who are part of a network to support mentorship across institutions. Participation in training such as CIMER's by CSB researchers and team leaders can be incentivized to grow an inclusive team environment where time spent learning skills to effectively manage mentoring relationships is valued by all stakeholders. Participation can be encouraged by building it into existing tenure and promotion requirements or other departmental career advancement and reward structures.

Addressing diversity, equity, inclusion, and accessibility (DEIA) in CSB necessitates multifaceted approaches. Initiatives such as BRC-BIO and ESIIL can offer targeted support to tackle specific challenges hindering DEIA in CSB. Moreover, addressing financial barriers by appropriately funding CSB training and professions, particularly within the biotechnology workforce, can help foster inclusivity. Beyond mere land acknowledgments, efforts could focus on training CSB leadership to effectively collaborate with Indigenous communities, with examples such as NEON Tribal Liaison positions serving as models for such endeavors. Finally, supporting training programs that promote systemic and cultural shifts can help develop supportive environments conducive to fostering DEIA in CSB.

A PATH FORWARD FOR TRAINING AND CAPACITY BUILDING IN CONTINENTAL-SCALE BIOLOGY

Focusing on data literacy, interdisciplinary team science, and promoting DEI is paramount for the effective development of CSB. The necessity of data literacy, ranging from basic to advanced expertise, is emphasized given the reliance on large-scale spatial and temporal data in CSB research. Additionally, the interdisciplinary nature of CSB underscores the importance of effective communication and collaboration across diverse backgrounds to ensure successful outcomes.

Furthermore, inclusivity and diverse perspectives are vital for maximizing creativity and productivity in CSB team science. Although training efforts exist in these areas, a more cohesive approach could weave these areas together effectively. This can be achieved through enhanced capacity building to aid in the sustainable overall support for research activities in CSB.

Moving forward, the following recommendation aims to guide not only the research community but also funders and decision makers in fostering a well-trained workforce equipped to address the multifaceted challenges of CSB. By prioritizing comprehensive training and capacity-building initiatives, we can better prepare a task force capable of conducting impactful research at various scales to respond effectively to global environmental crises.

Recommendation 5-1: The three key areas of training that funders, researchers, and educators should prioritize for developing a scientific workforce with the knowledge and skills necessary to address future challenges in CSB are data literacy, interdisciplinary team science, and promoting diversity, equity, inclusion, and accessibility.

REFERENCES

AECOM. 2022. Delivering today and more tomorrow. Annual report. AECOM Dallas, TX

Bennett, L.M., and H. Gadlin. 2012. Collaboration and team science: From theory to practice. *Journal of Investigative Medicine* 60:768-775. doi:10.231/JIM.0b013e318250871d.

Boillat, S., A. Martin, T. Adams, D. Daniel, J. Llopis, E. Zepharovich, C. Oberlack, G. Sonderegger, P. Bottazzi, E. Corbera, C. Ifejika Speranza, and U. Pascual. 2020. Why telecoupling research needs to account for environmental justice. *Journal of Land Use Science* 15:1-10. https://doi.org/10.1080/1747423X.2020.1737257.

Brockman, J.L., A.A. Nunez, and A. Basu. 2010. Effectiveness of a conflict resolution training program in changing graduate students style of managing conflict with their faculty advisors. *Innovative Higher Education* 35:277-293. https://doi.org/10.1007/s10755-010-9142-z.

Carroll, S.R., E. Herczog, M. Hudson, K. Russell, and S. Stall. 2021. Operationalizing the CARE and FAIR principles for Indigenous data futures. *Scientific Data* 8:108. https://doi.org/10.1038/s41597-021-00892-0.

Cheruvelil, K.S., and P.A. Soranno. 2018. Data-intensive ecological research is catalyzed by open science and team science. *BioScience* 68:813-822. https://doi.org/10.1093/biosci/biy097.

Cheruvelil, K.S., P.A. Soranno, K.C. Weathers, P.C. Hanson, S.J. Goring, C.T. Filstrup, and E.K. Read. 2014. Creating and maintaining high-performing collaborative research teams: The importance of diversity and interpersonal skills. *Frontiers in Ecology and the Environment* 12:31-38. https://doi.org/10.1890/130001.

Cummings, J.N., and S. Kiesler. 2007. Coordination costs and project outcomes in multi-university collaborations. *Research Policy* 36:1620-1634. http://dx.doi.org/10.1016/j.respol.2007.09.001.

Daehn, I.S., and P.L. Croxson. 2021. Disability innovation strengthens STEM. *Science* 373:1097-1099. https://doi.org/10.1126/science.abk263.

Demery, AJ.C., and M.A. Pipkin. 2021. Safe fieldwork strategies for at-risk individuals, their supervisors and institutions. *Nature Ecology & Evolution* 5:5-9. https://doi.org/10.1038/s41559-020-01328-5.

Dodds, W.K., K.C. Rose, S. Fei, and S. Chandra. 2021. Macrosystems revisited: challenges and successes in a new subdiscipline of ecology. *Frontiers in Ecology and the Environment* 19:4-10. doi:10.1002/fee.2286.

Elliott, K.C., K.S. Cheruvelil, G.M. Montgomery, and P.A. Soranno. 2016. Conceptions of good science in our data-rich world. *BioScience* 66:880-889. https://doi.org/10.1093/biosci/biw115.

Elmqvist, T., E. Andersson, and T. McPhearson, X. Bai, L. Bettencourt, E. Brondizio, J. Colding, G. Daily, C. Folke, N. Grimm, D. Haase, D. Ospina, S. Parnell, S. Polasky, K. C. Seto, and S. Van Der Leeuw. 2021. Urbanization in and for the Anthropocene. *NPJ Urban Sustain* 1:6. https://doi.org/10.1038/s42949-021-00018-w.

Farley, S.S., A. Dawson, S.J. Goring, and J.W. Williams. 2018. Situating ecology as a big-data science: Current advances, challenges, and solutions. *BioScience* 68:563-576. https://doi.org/10.1093/biosci/biy068.

Fiore, S.M. 2008. Interdisciplinarity as teamwork. How the science of teams can inform team science. *Small Group Research* 39:251-277. https://doi.org/10.1177/1046496408317797.

Fiore, S.M., J. Johnston, and R. McDaniel. 2005. Applying the narrative form and XML metadata to debriefing simulation-based exercises. Pp. 2135-2139 in *Proceedings of the 49th Annual Meeting of the Human Factors and Ergonomics Society*. Santa Monica, CA: Human Factors and Ergonomics Society.

Fortunato, S., C.T. Bergstrom, J.A. Evans, J.A. Evans, D. Helbing, S. Milojević, A.M. Petersen, F. Radicchi, R. Sinatra, B. Uzzi, A. Vespignani, L. Waltman, D. Wang, and A.-L. Barabási. 2018. Science of science. *Science* 359:eaao0185(2018). doi:10.1126/science.aao0185.

Grantham, H.S. 2022. Forest conservation: Importance of Indigenous lands. *Current Biology* 32:R1262-R1286. https://doi.org/10.1016/j.cub.2022.10.026.

Hall, K.A. Vogel, G. Huang, K. Serrano, E. Rice, S. Tsakraklides, and S. Fiore. 2018. The science of team science: A review of the empirical evidence and research gaps on collaboration in science. *American Psychologist* 73:532-548. https://doi.org/10.1037/amp0000319.

Hampton, S.E., C.A. Strasser, J.J. Tewksbury, W.K. Gram, A.E. Budden, A.L. Batcheller, C.S. Duke, and J.H. Porter. 2013. Big data and the future of ecology. *Frontiers in Ecology & the Environment* 11:156-162. https://doi.org/10.1890/120103.

Hampton, S.E., M.B. Jones, L.A. Wasser, M.P. Schildhauer, S.R. Supp, J. Brun, R.R. Hernandez, C. Boettiger, S.L. Collins, L.J. Gross, D.S. Fernández, A. Budden, E.P. White, T.K. Teal, S.G. Labou, and J.E. Aukema. 2017. Skills and knowledge for data-intensive environmental research. *BioScience* 67:546-557. https://doi.org/10.1093/biosci/bix025.

Hofstra, B., V.V. Kulkarni, S. Galvez Sebastian Munoz-Najar, B. He, D. Jurafsky, and D.A. McFarland. 2020. The diversity-innovation paradox in science. *Proceedings of the National Academy of Sciences of the United States of America* 117:9284-9291.

IOM (Institute of Medicine). 2005. *Facilitating Interdisciplinary Research*. Washington, DC: The National Academies Press. https://doi.org/10.17226/11153.

Jensen, A.J., S.P. Bombaci, L.C. Gigliotti, S.N. Harris, C.J. Marneweck, M.S. Muthersbaugh, B.A. Newman, S.L. Rodriguez, E.A. Saldo, K.E. Shute, K.L. Titus, A.L. Williams, S.W. Yu, and D.S. Jachowski. 2021. Attracting diverse students to field experiences requires adequate pay, flexibility, and inclusion. *BioScience* 71:757-770 https://doi.org/10.1093/biosci/biab039.

Kjelvik, M.K., and E.H. Schultheis. 2019. Getting messy with authentic data: Exploring the potential of using data from scientific research to support student data literacy. *CBE—Life Sciences Education* 18(2). https://www.lifescied.org/doi/full/10.1187/cbe.18-02-0023.

Klomparens, K., J. Beck, J. Brockman, and A. Nunez. 2008. *Setting Expectations and Resolving Conflicts in Graduate Education*. Washington, DC: Council of Graduate Schools.

Langen, T.A., T. Mourad, B. Grant, W.K. Gram, B.J. Abraham, D.S. Fernandez, M. Carroll, A. Nuding, J.K. Balch, J. Rodriguez, and S.E. Hampton. 2014. Using large public datasets in the undergraduate ecology classroom. *Frontiers in Ecology & the Environment* 12:362-363. https://doi.org/10.1890/1540-9295-12.6.362.

Lundin, M., and S. Bombaci. 2022. Making outdoor field experiences more inclusive for the LGBTQ+ community. *Ecological Applications* 33(5):e2771. https://doi.org/10.1002/eap.2771.

NASEM (National Academies of Sciences, Engineering, and Medicine). 2021. *Call to Action for Science Education: Building Opportunity for the Future*. Washington, DC: The National Academies Press. https://doi.org/10.17226/26152.

NASEM. 2023. *Transforming EPA Science to Meet Today's and Tomorrow's Challenges*. Washington, DC: The National Academies Press. https://doi.org/10.17226/26602.

NASEM. 2024. *Disrupting Ableism and Advancing STEM: Promoting the Success of People with Disabilities in the STEM Workforce: Proceedings of a Workshop Series*. Washington, DC: The National Academies Press. https://doi.org/10.17226/27245.

National Council of Nonprofits. 2022. Nonprofit impact matters: How America's charitable nonprofits strengthen communities and improve lives. National Council of Nonprofits. Washington, DC.

NRC (National Research Council). 2004. *NEON: Addressing the Nation's Environmental Challenges*. Washington, DC: The National Academies Press. https://doi.org/10.17226/10807.

NRC. 2015. *Enhancing the Effectiveness of Team Science*. Washington, DC: The National Academies Press. https://doi.org/10.17226/19007.

National Science and Technology Council Environmental Justice Subcommittee. 2024. *Environmental Justice, Data, and Research Plan*. White House Office of Science and Technology Policy, Washington, DC.

Newell, W.H., and D.B. Luckie. 2013. Pedagogy for interdisciplinary habits of the mind. In *Conference on Interdisciplinary Teaching and Learning*, A.M. McCright and W. Eaton, eds. May 2012 White Paper. East Lansing, MI: Michigan State University.

Nersessian, N.J. 1999. Model-based reasoning in conceptual change. Pp. 5-22 in *Model-Based Reasoning in Scientific Discovery*, L. Magnani, N.J. Nersessian, and P. Thagard, eds. Springer.

Nielsen, M.W., A. Sharla, B. Love, H. Etzkowitz, H.J. Falk-Krzesinski, A. Joshi, E. Leahey, L. Smith-Doerr, A.W. Woolley, and L. Schiebinger. 2017. Gender diversity leads to better science. *Proceedings of the National Academy of Sciences of the United States of America* 114:1740-1742. https://doi.org/10.1073/pnas.1700616114.

Pennington, D., G. Simpson, M. McConnell, J. Fair, and R. Baker. 2013. Transdisciplinary science, transformative learning, and transformative science. *BioScience* 63:564-573. https://doi.org/10.1525/bio.2013.63.7.9.

Pennington, D., I. Ebert-Uphoff, N. Freed, J. Martin, and S. A. Pierce. 2020. Bridging sustainability science, Earth science, and data science through interdisciplinary education. *Sustainability Science* 15:647-661. https://doi.org/10.1007/s11625-019-00735-3.

Peterson, D.M., S.M. Flynn, R.S. Lanfear, C. Smith, L.J. Swenson, A.M. Belskis, S.C. Cook, C.T. Wheeler, J.F. Wilhelm, and A.J. Burgin. 2023. Team science: A syllabus for success on big projects. *Ecology and Evolution* 13:e10343. https://doi.org/10.1002/ece3.10343.

Read, E.K., M. O'Rourke, G.S. Hong, P.C. Hanson, L.A. Winslow, S. Crowley, C.A. Brewer, and K.C. Weathers. 2016. Building the team for team science. *Ecosphere* 7:e01291. https://doi.org/10.1002/ecs2.1291.

Stagl, K.C., E. Salas, and S.M. Fiore. 2007. Best practices in cross training teams. Pp. 155-179 in *Workforce Cross Training Handbook*, D.A. Nembhard, ed. Boca Raton, FL: CRC Press.

Sterner, B., and S. Elliott, S. 2022. The FAIR and CARE data principles influence who counts as a participant in biodiversity science by governing the fitness-for-use of data. philsci-archive.pitt.edu/21039/1/FAIRandCAREData-April6-2022.pdf (accessed May 2, 2024).

Stewart, A.J. 2021. Dismantling structural racism in academic psychiatry to achieve workforce diversity. *American Journal of Psychiatry* 178:210-212. https://doi.org/10.1176/appi.ajp.2020.21010025.

UNESCO (United Nations Educational, Scientific and Cultural Organization). 2006. *Education for All: Literacy for Life*. Global Education Monitoring Report. Paris: UNESCO https://doi.org/10.54676/HFRH4626.

Vacca, R., C. McCarty, M. Conlon, and D.R. Nelson. 2015. Designing a CTSA-based social network intervention to foster cross-disciplinary team science. *Clinical and Translational Science* 8:281-289. https://doi.org/10.1111/cts.12267.

Videnovik M., E. Vlahu-Gjorgievska, and V. Trajkovik. 2021. To code or not to code: Introducing coding in primary schools. *Computer Applications in Engineering Education* 29:1132-1145. https://doi.org/10.1002/cae.22369.

White House. 2021a. Executive Order on Diversity, Equity, Inclusion, and Accessibility in the Federal Workforce. https://www.whitehouse.gov/briefing-room/presidential-actions/2021/06/25/executive-order-on-diversity-equity-inclusion-and-accessibility-in-the-federal-workforce/.

White House. 2021b. White House Commits to Elevating Indigenous Knowledge in Federal Policy Decisions. https://www.whitehouse.gov/ostp/news-updates/2021/11/15/white-house-commits-to-elevating-indigenous-knowledge-in-federal-policy-decisions/.

Wiek, A., M.J. Bernstein, R.W. Foley, M. Cohen, N. Forrest, C. Kuzdas, and L.W. Keeler. 2015. Operationalising competencies in higher education for sustainable development. Pp. 241-260 in *Routledge Handbook of Higher Education for Sustainable Development,* M. Barth, G. Michelsen, M. Rieckmann, and I. Thomas, eds. London: Routledge.

Wilson, G., D.A. Aruliah, C.T. Brown, C.T. Brown, N.P.C. Hong, M. Davis, R.T. Guy, S.H.D. Haddock, K.D. Huff, I.M. Mitchell, M.D. Plumbley, B. Waugh, E.P. White, and P. Wilson. 2014. Best practices for scientific computing. *PLoS Biology* 12(1):e1001745. https://doi.org/10.1371/journal.pbio.1001745.

Xia, L., and S. Ya. 2012. Study on knowledge sharing behavior engineering. *Systems Engineering Procedia* 4:468-476. https://doi.org/10.1016/j.sepro.2012.01.012.

Zhu, K., P. Das, A. Karimuddin S. Tiwana, J. Siddiqi, and F. Khosa. 2021. Equity, diversity, and inclusion in academic American surgery faculty: An elusive dream. *Journal of Surgical Research* 258:179-186. https://doi.org/10.1016/j.jss.2020.08.069.

6

Overarching Recommendations and Vision

OVERARCHING RECOMMENDATIONS

Continental-scale biology (CSB) is built on connections: from teleconnections (causal links between phenomena in geographically distant regions) to feedbacks between ecosystems and ecosystem components, to cross-scale interactions that occur when phenomena at one spatial scale influence another. In addition, virtually every natural system on Earth influences and is influenced by human activities, even over long distances. Bringing together these factors is a central challenge of CSB research. Here, the committee details two overarching recommendations that will help to meet this challenge and support the development of the emerging field of CSB.

As detailed in Chapter 1, several recent National Science Foundation (NSF) initiatives have sought to enhance understanding of biological systems by integrating the methods and knowledge from the many scientific subdisciplines and at many different scales. These programs include Reintegrating Biology, Understanding the Rules of Life, the Biological Integration Institutes, and Macrosystems Biology (see Chapter 1, Box 1-1). However, the scope of CSB is broader than that of any of the existing NSF programs, and CSB research has special complex logistical challenges in many aspects such as data sharing and cyberinfrastructure needs across multiple scales.

For example, the core programs in the Division of Environmental Biology support "research and training on evolutionary and ecological processes acting at the level of populations, species, communities, and ecosystems"[1] but CSB also addresses processes below population levels (e.g., subcellular, cellular) and above ecosystem levels (e.g., regional, continental). Even if the purposes of these core programs are expanded, it would be challenging for each of the core programs to efficiently and effectively

---

[1] See https://new.nsf.gov/funding/opportunities/division-environmental-biology-deb?utm_medium=email&utm_source=govdelivery (accessed 5 March 2024).

evaluate CSB proposals because each core program would need to have a much wider range of experts to evaluate CSB proposals. Therefore, CSB would strongly benefit from the establishment of an NSF core program that provides a stable and dedicated funding source to support research addressing the interplay of organizational, temporal, and spatial scales and is based on integrated yet flexible frameworks from a systems perspective.

Overarching Recommendation 1: The National Science Foundation should establish a core program on CSB.

The committee recommends that NSF establish a new core program on CSB. This could be a joint effort among the relevant NSF divisions and directorates to help facilitate collaborations, both between the Division of Environmental Biology and other divisions within the Directorate for Biological Sciences (e.g., the Division of Integrative and Organismal Systems and the Division of Molecular and Cellular Biosciences), and with other directorates, such as Mathematical and Physical Sciences, Computer and Information Science and Engineering, Office of Advanced Cyberinfrastructure, Engineering, and Social, Behavioral, and Economic Sciences. For example, collaboration with the Directorate for Technology, Innovation, and Partnerships could help with the development of new technologies that would advance CSB; work with the Directorate for Social, Behavioral, and Economic Sciences would help provide additional insight on the increasing influence of human activities on biological systems and, conversely, the effects of biological systems on human well-being; and collaboration with the Directorate for Geosciences would support work on the linkages between geophysical and atmospheric processes and CSB.

Overarching Recommendation 2: Researchers and funders should develop CSB under integrated yet flexible frameworks.

Second, as described in Chapter 1 and discussed in more detail in the connectivity theme in Chapter 2, CSB addresses questions about biological processes and patterns that emerge at broad organizational, spatial, and/or temporal scales and treats biological systems as part of coupled human and natural systems, given widespread human impacts and intensifying human–nature interactions worldwide. Integrated but flexible frameworks for CSB would enable researchers to better understand and contextualize the connections between data inputs from the biological, abiotic, and socioeconomic realms, and the interactions within, between, and among adjacent and distant locations. Such frameworks would help researchers gain a holistic view of local- and regional-scale ecosystems and continental-scale environmental shifts—insights that will allow the development of more effective and sustainable solutions to the environmental and ecological challenges facing our planet.

An example of an integrated framework could be based on the metacoupling framework (see Box 2-4). The metacoupling framework has been applied to analyses of many topics, including ecosystem services, resilience, vulnerability, biodiversity conservation, biogeochemical flows, climate change, freshwater use, land use, pollution, impacts of food imports on food security, and effects of international trade on deforestation. Its applications have been reported in terrestrial, freshwater, coastal, and marine ecosystems in locations encompassing the Antarctic, tropical, temperate, and Arctic regions; urban and rural areas; and upstream, midstream, and downstream regions (Liu 2023). Further, the metacoupling framework expands and integrates many existing concepts, theories, and disciplines. Examples include general systems theory, ecology, geography, metapopulation, metacommunity, meta-ecosystems, scale, teleconnection, and ecosystem services. Although the framework is comprehensive, it is also flexible. For example, a given analysis need not address all the components of the framework but placement under the framework allows it to be linked with other analyses that use the framework, enhancing the collective impact.

VISION

CSB will help scientists address challenges such as understanding how fundamental life processes on Earth are changing across various scales, with respect to increasing human domination in ecosystems all over the planet. As global change and human impacts intensify, Earth's fundamental processes—from atmospheric circulation to the cycling of nutrients—will experience significant shifts. CSB can enable scientists to integrate data from global-scale observations with data gathered at regional and finer scales to help capture evidence of these shifts. These data will contribute to the development of theory, models, and mechanistic insights, in turn making it possible to identify patterns and processes and determine how they are changing.

CSB can also help address questions around ecosystem services, migration, disease spread, gene flow, and evolution. From a continental-scale viewpoint, scientists will be able to answer questions about how continent-wide environmental and land-use patterns are shaping regional and local-scale changes in ecosystem services, how different ecosystems are responding to invasive plant and animal species, and how infectious diseases arrive at new locations.

CSB can also help to inform and support the conservation of global biodiversity and ecosystem services. By deciphering biological processes from cells to ecosystems, CSB can help identify priority areas for restoration and conservation, for example, by identifying areas that harbor distinct biodiversity, essential corridors for migrations, or regions that allow range-shifting species to adjust to climate and land-use change. Similarly, CSB can help identify which species, populations, and ecosystems are more vulnerable or more resilient to global changes, information that could help inform policy and management decisions. CSB could also help identify nature-based solutions that can benefit biodiversity and ecosystem conservation as well as mitigating the effects of climate change.

CSB can help human health and survival. A major human–nature interaction is agriculture, which is central to food systems for human health and survival (Barrett et al. 2023, Fanzo and Miachon 2023, Sanchez 2020). Agriculture is closely related to the themes of CSB, as it influences and is influenced by biodiversity and ecosystem functions, resilience and vulnerability, connectivity, and sustainability. For example, agriculture uses a large proportion of land area, accounts for the majority of human-consumed freshwater, generates a significant amount of carbon, affects biodiversity through land conversion and applications of agricultural chemicals such as fertilizers and pesticides, and connects various parts of the world through trade of agricultural chemicals and food (Hanemann and Young 2020, Heal 2020, Kling et al. 2017). On the other hand, biodiversity is essential for agriculture by providing crop varieties and pollination (Dasgupta 2021, Heal 2020). To maximize the benefits and minimize the negative effects, it is essential to implement effective cross-scale planning, governance, and management (Brown et al. 2021, Lemos and Agrawal 2006, McCay et al. 2014, Segerson 2022).

CSB also opens the door for greater interagency collaboration among various agencies. For example, as agriculture is closely related to the CSB themes, it is logical for the U.S. Department of Agriculture to be an important partner to support and benefit from CSB, just like the great work that has occurred through cross-agency collaboration on plant genomes.

Bold initiatives are needed to create a truly continental-scale biology that addresses questions about complex biological systems including human and abiotic factors across multiple, organizational, spatial, and temporal scales from a systems perspective. The vision for a new era of CSB is one that integrates across biological subdisciplines as well as other disciplines and across multiple scales of research, harnessing the power of the biological data revolution to address questions that cannot be answered by observations and experiments conducted at either fine or large scales alone.

Although much progress has been made, there are many major gaps in knowledge, theory, data, networks, tools, and training and capacity building needed to support the vision for CSB. Filling these gaps will require the development of new theories and technologies that encompass not just biology, but atmospheric sciences, mathematics, engineering, physics, geosciences, environmental chemistry, and social sciences. Such effort is crucial to enhance fundamental understanding of ongoing changes in biodiversity, ecosystem services, climate, disease spread, species invasion, gene flows, and biotic interactions. It is also needed to build workforce capacity by mentoring a new generation of innovative scholars and engaging leaders for global sustainability. By addressing these challenges with coordinated and innovative efforts, we can pave the way for a sustainable and resilient future, ensuring the well-being of our planet and its ecosystems.

REFERENCES

Barrett, C.B., A. Ortiz-Bobea, and T. Pham 2023. Structural transformation, agriculture, climate, and the environment. *Review of Environmental Economics and Policy* 17:195-216. https://doi.org/10.1086/725319.

Brown, M.A., P. Dwivedi, S. Mani, D. Matisoff, J.E. Mohan, J. Mullen, M. Oxman, M. Rodgers, R. Simmons, B. Beasley, and L. Polepeddi. 2021. A framework for localizing global climate solutions and their carbon reduction potential. *Proceedings of the National Academy of Sciences of the United States of America* 118(31):e2100008118. https://doi.org/10.1073/pnas.2100008118.

Dasgupta, P. 2021. *The Economics of Biodiversity: The Dasgupta Review*. London: HM Treasury.

Fanzo, J., and L. Miachon. 2023. Harnessing the connectivity of climate change, food systems and diets: Taking action to improve human and planetary health. *Anthropocene* 42:100381. https://doi.org/10.1016/j.ancene.2023.100381.

Hanemann, M., and M. Young. 2020. Water rights reform and water marketing: Australia vs the US West. *Oxford Review of Economic Policy* 36:108-131. https://doi.org/10.1093/oxrep/grz037.

Heal, G. 2020. *The Economic Case for Protecting Biodiversity*. Working Paper 27963. National Bureau of Economic Research. https//doi.org/10.3386/w27963.

Kling, C.L., R.W. Arritt, G. Calhoun, and D.A. Keiser. 2017. Integrated assessment models of the food, energy, and water nexus: A review and an outline of research needs. *Annual Review of Resource Economics* 9(1):143-163. https://doi.org/10.1146/annurev-resource-100516-033533.

Lemos, M.C., and A. Agrawal 2006 Environmental governance. *Annual Review of Environment and Resources* 31:297-325. https://doi.org/10.1146/annurev.energy.31.042605.135621.

Liu, J. 2023. Leveraging the metacoupling framework for sustainability science and global sustainable development. *National Science Review* 10(7):nwad090. https://doi.org/10.1093/nsr/nwad090.

McCay, B.J., F. Micheli, G. Ponce-Díaz, G. Murray, G. Shester, S.I. Ramirez-Sanchez, and W. Weisman. 2014. Cooperatives, concessions, and co-management on the Pacific coast of Mexico. *Marine Policy* 44:49-59. https://doi.org/10.1016/j.marpol.2013.08.001.

Sanchez, P.A. 2020. Time to increase production of nutrient-rich foods. *Food Policy* 91:101843. https://doi.org/10.1016/ j.foodpol.2020.101843.

Segerson, K. 2022. Group incentives for environmental protection and natural resource management. *Annual Review of Resource Economics* 14:597-619. https://doi.org/10.1146/annurev-resource-111920-020235.

Appendix A

Committee Member Biographical Sketches

Jianguo Liu (*Chair*) holds the Rachel Carson Chair in Sustainability, is University Distinguished Professor at Michigan State University, and serves as director of the Center for Systems Integration and Sustainability. He has previously served as a visiting scholar at Harvard, Princeton, and Stanford universities. An ecologist, human–environment scientist, and sustainability scholar, Liu takes a holistic approach to addressing complex human–environmental challenges across local to global scales through systems integration, such as the integration of ecology with social sciences, policy, and technologies. He is particularly keen to connect seemingly unconnected issues, for example, telecoupling (human–nature interactions over distances), divorce, and environmental sustainability. Liu is a former president of the North American Regional Association of the International Association for Landscape Ecology. He has served on various international and national committees, panels, and editorial boards of international journals such as *Science*. He has received numerous awards, including the Eminent Ecologist Award of the Ecological Society of America, the Gunnerus Award in Sustainability Science from the Royal Norwegian Society of Sciences and Letters and the Norwegian University of Science and Technology, and the World Sustainability Award from the MDPI Sustainability Foundation in Switzerland. He is an elected member of the American Philosophical Society, American Academy of Arts and Sciences, and Royal Norwegian Society of Sciences and Letters. Liu received a Ph.D. in ecology from the University of Georgia and completed his postdoctoral study at Harvard University.

Jeannine Cavender-Bares is Distinguished McKnight Professor at the University of Minnesota in the Department of Ecology, Evolution and Behavior and the director of the National Science Foundation Biology Integration Institute ASCEND (Advancing Spectral biology in Change ENvironments to understand Diversity). ASCEND seeks to understand the causes and consequences of plant biodiversity across spatial and biological scales using reflected photons and process-based predictive models. Her work focuses on the physiological and evolutionary dimensions of plant ecology that influence community assembly and ecosystem function. Cavender-Bares is particularly interested in the genetic and evolutionary basis of variation in plant phenotypes and spectral properties that can advance remote sensing of biodiversity. She is on the Governing Board of the Ecological Society of America and a member of the American Academy of Arts and Sciences. Cavender-Bares received a Ph.D. in organismic and evolutionary biology from Harvard University.

Bala Chaudhary is an associate professor of environmental studies and the Ecology, Evolution, Environment and Society Program at Dartmouth College. She previously held faculty appointments at DePaul University and Loyola University Chicago. Chaudhary's primary research expertise is in plant-soil-microbial ecology spanning organizational scales from genes to ecosystems across spatial scales. Research in her lab uses continent-wide field experiments, big data synthesis, and trait-based approaches to study fungal dispersal at macrosystem scales. Prior to academia, Chaudhary worked as a restoration consultant designing habitat mitigation plans for endangered species management. Chaudhary is a National Science Foundation CAREER awardee, committee member for the Ecological Society of America, and award-winning advocate for diversity, equity, and inclusion in STEM. She earned a B.A. in biological sciences from the University of Chicago and an M.S. and a Ph.D. in biological sciences from Northern Arizona University.

Brian J. Enquist is a professor at the University of Arizona and an external faculty member of the Santa Fe Institute. He is a broadly trained physiologist, ecologist, and botanist whose research program investigates the origin and maintenance of biological diversity and the functioning of the biosphere. His lab uses biological scaling laws and is developing a general trait-based theory. His lab strives to develop a more integrative, quantitative, and predictive framework for biology, community ecology, and global ecology. Enquist is a fellow of the Ecological Society of America (ESA) and the American Association for the Advancement of Science. He is a Fulbright Scholar (Costa Rica), a National Geographic Explorer, and a recipient of the ESA's Mercer Award, and a National Science Foundation (NSF) CAREER Award. He was a fellow at the Center for Theoretical Study in Prague, Czech Republic; the Centre national de la recherche scientifique (CNRS) in Montpellier, France; and the Oxford Martin School at Oxford University in the United Kingdom. Enquist received a Ph.D. in biology in 1998 from the University of New Mexico and was an NSF postdoctoral fellow at the National Center for Ecological Analysis and Synthesis at the University of California, Santa Barbara.

Jack A. Gilbert is a professor in pediatrics and the Scripps Institution of Oceanography and associate vice chancellor for marine science at the University of California, San Diego (UCSD). Prior to his current position, he was group leader for Microbial Ecology at Argonne National Laboratory, a professor of surgery, and director of the Microbiome Center at the University of Chicago. He cofounded the Earth Microbiome Project and

American Gut Project. He is the founding editor in chief of *mSystems* journal. Gilbert co-authored *Dirt is Good,* a popular science guide to the microbiome and children's health and founded BiomeSense, Inc. to produce automated microbiome sensors. In 2021, Gilbert became the principal investigator and director of the UCSD Microbiome and Metagenomics Center. In 2023, he was elected to president of Applied Microbiology International. Gilbert received a Ph.D. in molecular microbiology from the University of Nottingham.

N. Louise Glass is a recent emeritus professor (2023) in the Plant and Microbial Biology Department (PMB) at the University of California, Berkeley, and a senior faculty scientist in the Environmental Genomics and Systems Biology Division at the Lawrence Berkeley National Laboratory (LBNL). Glass was previously chair/associate chair of the PMB Department and previously was director of the Environmental Genomics and Systems Biology Division at LBNL. Her research focus has been on filamentous fungi, exploring molecular mechanisms of cell-to-cell communication and somatic cell fusion associated with hyphal network formation, plant cell wall deconstruction by fungi and how these processes affect ecosystems. Her recent work has focused on developing high-throughput functional genomics for nonmodel fungi with an aim to understanding and controlling the interactions between plant roots and neighboring microbes to gain insights into carbon cycling, carbon sequestration, and plant productivity in natural and agricultural ecosystems. Glass is the Fred E. Dickinson Chair of Wood Science and Technology; a fellow of the American Society for the Advancement of Science, the American Academy of Microbiology, and the Mycological Society of America; a recipient of an Alexander von Humboldt Research Award; and was elected to the National Academy of Sciences in 2021. Glass received a Ph.D. in plant pathology from the University of California, Davis, and performed postdoctoral work on fungal genetics and molecular biology at the University of Wisconsin–Madison.

Scott Goetz is a Regents Professor of Earth Observation & Ecological Informatics at Northern Arizona University. He has conducted satellite remote sensing research over the past 30+ years. He has served on working groups for the Intergovernmental Panel on Climate Change; United Nations programs on Reducing Emissions from Deforestation and Forest Degradation (REDD+); the U.S. Global Change Research Program; the National Academies of Sciences, Engineering, and Medicine; and interagency programs on carbon cycle science, climate change, and terrestrial ecology. He is science lead of NASA's Arctic Boreal Vulnerability Experiment and deputy principal investigator of NASA's Global Ecosystem Dynamics Investigation. He has mentored dozens of early-career scientists and graduate students, is editor in chief of *Environmental Research Ecology,* executive board member of *Environmental Research Letters,* past deputy director of the Woods Hole Research Center, a Fulbright Research Scholar (France), and past associate editor of the *Journal of Geophysical Research: Biogeosciences* and *Remote Sensing of Environment.* Goetz received a Ph.D. in geographic sciences from the University of Maryland, College Park.

Stephanie E. Hampton is the deputy director of Biosphere Sciences and Engineering at the Carnegie Institution for Science, a division that aims to integrate biology from molecular to global scales. Prior to 2022, she served at the National Science Foundation (NSF) as director for the Division of Environmental Biology. While at NSF, she remained a professor in the School of the Environment at Washington State University (WSU). Prior to WSU, she was deputy director of the National Center for Ecological Analysis and Synthesis at the University of California, Santa Barbara. Hampton is a freshwater ecologist with expertise in analysis of large environmental datasets, and her research includes analyzing the effects of climate change and other human impacts on lakes worldwide. She received the 2020 Ramón Margalef Award for Excellence in Education from the Association for the Sciences of Limnology and Oceanography for training and mentoring in data-intensive research, and the 2017 Chandler-Misener Award from International Association for Great Lakes Research with her American–Russian team working on Lake Baikal in Siberia. She is 2023 president-elect for the Ecological Society of America. Hampton received a B.A. in environmental studies from the University of Kansas, an M.S. in biology from the University of Nevada, Las Vegas, and a Ph.D. in ecology and evolution from Dartmouth College.

Inés Ibáñez is a professor in global change ecology in the School for Environment and Sustainability at the University of Michigan. Her research group works on the effects of global change factors on terrestrial ecosystems, focusing on forecasting resilience and vulnerability of plant communities to climate change, pollution, introduced species, and landscape degradation. Her work expands from the physiological and demographic performance of tree species to landscape dynamics and continental-scale patterns of ecosystem responses to global change. Her computational work centers on developing integrated models of forest multifunctionality that incorporate processes, products, and drivers across the atmospheric, vegetation, and soil components of a forest with the goal of identifying connections, feedbacks, and thresholds that advance integrated science and that inform sustainable management and conservation. Ibáñez received a B.S. in biology from Universidad Complutense, Spain, an M.S. in ecology from Utah State University, and a Ph.D. in ecology from Duke University. She was a member of the National Academies of Sciences, Engineering, and Medicine's Committee on the Potential for Biotechnology to Address Forest Health.

Chelcy F. Miniat is a biological scientist and research program manager with the USDA Forest Service, Rocky Mountain Research Station, Maintaining Resilient Dryland Ecosystems science program. Her program spans six states in the Intermountain Western United States. Research in her program is focused on grassland, shrubland, and desert ecosystems, which cover approximately 900 million acres across 17 U.S. western states. Her program's primary areas of emphasis are (1) understanding disturbances and stressors, reducing risk, and increasing resilience and resistance; (2) developing restoration tools, guidelines, and applications to restore ecosystems and control invasive species; and (3) understanding the effects of climate change and climate variability at multiple scales. She served a significant role in writing the Forestry Sector Report and the Climate and Fire technical input reports for the 2013 National Climate Assessment, the 2015 USDA Forest Service Drought Synthesis, led significant portions of the 2021 USDA Invasive Species National Assessment, and worked on the Fifth National Climate Assessment Forestry Chapter. Ford Miniat received a B.S. in applied biology from the Georgia Institute of Technology, an M.S. in botany from the University of South Florida, and a Ph.D. in forest resources from the University of Georgia.

Shahid Naeem is a professor of ecology and chair of the Department of Ecology, Evolution, and Environmental Biology at Columbia University. His research has focused on the ecological and environmental consequences of biodiversity loss across all scales. His theoretical, experimental, and observational studies have been conducted across a wide array of organisms (plants, animals and microorganisms), habitats (freshwater, terrestrial, and marine), and ecosystems (tundra, rainforests, grasslands, urban- and agro-ecosystems) in countries around the world. He is an Aldo Leopold Leadership fellow; a fellow of the American Association for the Advancement of Science, the Ecological Society of America (ESA), and Michigan Society of Fellows; was elected president of ESA in 2022; and is the recipient of the Buell and Mercer Awards from the ESA and Lenfest Award from Columbia. Naeem received a Ph.D. in zoology from the University of California, Berkeley, and served as a postdoctoral fellow at the University of Michigan, Imperial College of London, and the University of Copenhagen.

Phoebe L. Zarnetske is a professor of spatial and community ecology in the Department of Integrative Biology at Michigan State University (MSU) and is director of the Institute for Biodiversity, Ecology, Evolution, and Macrosystems. Her research integrates insights from climate change experiments with macrosystems science and modeling of big data in ecology across scales. Her research on climate change ecology has elucidated important roles of biotic interactions among species, and how these interactions can exacerbate the impacts of climate change on biodiversity across scales. She co-leads the National Science Foundation (NSF)-funded Climate Intervention Biology Working Group, bringing together experts in climate science and ecology to research the potential ecological impacts from climate intervention. She received the MSU College of Natural Science Early Career Research Award, is lead principal investigator (PI) of NSF Macrosystems NEON and NASA grants, and is a co-PI of the Kellogg Biological Station Long Term Ecological Research (LTER) site. Zarnetske received a B.A. in biology from Colby College, an M.S. in ecology from Utah State University, and a Ph.D. in integrative biology from Oregon State University where she was an NSF Integrative Graduate Education and Research Traineeship fellow. She completed her postdoctoral training as a Yale Climate and Energy Institute postdoctoral fellow in the Yale School of the Environment. She is a member of the National Academies of Sciences, Engineering, and Medicine's workshop on Climate Intervention in an Earth Systems Science Framework.

Appendix B

Public Meeting Agendas

**COMMITTEE ON RESEARCH AT MULTIPLE SCALES:
A VISION FOR CONTINENTAL SCALE BIOLOGY**

MARCH 16, 2023

The first public meeting of the Committee on Research at Multiple Scales: A Vision for Continental Scale Biology was held virtually.

Open Session Agenda
March 16, 2023
2:00 p.m.–3:00 p.m.

2:00 **Welcome and Introductions**
Jianguo Liu, *Committee Chair, National Academies of Sciences, Engineering, and Medicine*
Clifford Duke, *Study Director, National Academies of Sciences, Engineering, and Medicine*

2:10 **Discussion of Study Statement of Task with Sponsor and Committee**
Matt Kane, *National Science Foundation*

4:30 **Adjourn Open Session**

APRIL 24–25, 2023

The second public meeting of the Committee on Research at Multiple Scales: A Vision for Continental Scale Biology was held virtually. This was the first of three information-gathering sessions held via a webinar series.

**Paving the Way for Continental Scale Biology:
Connecting Research Across Scales Agenda**
April 24, 2023
1:30 p.m.–4:00 p.m.

1:30 **Welcome**
Clifford Duke, *Study Director, National Academies of Sciences, Engineering, and Medicine*

1:35	**Opening Remarks**
	Jianguo Liu, *Committee Chair, National Academies of Sciences, Engineering, and Medicine*
1:45	**Keynote Speaker—Forecasting Global Change Impacts on Biodiversity**
	Janet Franklin, *San Diego State University*
2:30	**Panel—Biology at Multiple Scales from Researchers' Perspectives**
	David Schimel, *National Aeronautics and Space Administration*
	Emiley Eloe-Fadrosh, *Lawrence Berkeley National Laboratory*
	Sydne Record, *University of Maine*
	Marten Winter, *German Centre for Integrative Biodiversity Research (iDiv) Halle-Jena-Leipzig/Univ Leipzig*
	Noah Fierer, *University of Colorado Boulder*
	Thomas Elmqvist, *Stockholm University*
3:55	**Key Takeaways from Day 1**
	Jianguo Liu, *Committee Chair, National Academies of Sciences, Engineering, and Medicine*
4:00	**Adjourn Open Session**

April 25, 2023
1:00 p.m.–3:30 p.m.

1:00	**Welcome**
	Jianguo Liu, *Committee Chair, National Academies of Sciences, Engineering, and Medicine*
1:05	**Keynote Speaker—Ecosystems and the Biosphere as Complex Adaptive Systems: Scaling, Collective Phenomena, and Governance**
	Simon Levin, *Princeton University*
1:50	**Panel—Biology at Multiple Scales from Program Managers' Perspectives**
	Anika Dzierlenga, *National Institutes of Health*
	Katharina Dittmar, *National Science Foundation*
	Woody Turner, *National Aeronautics and Space Administration*
	Todd Anderson, *Department of Energy*
	Scott Hagerthey, *Environmental Protection Agency*
	Michael Wilson, *U.S. Department of Agriculture*

3:20   **Key Takeaways from Day 2**
Jianguo Liu, *Committee Chair, National Academies of Sciences, Engineering, and Medicine*

3:30   **Adjourn Open Session**

JUNE 15, 2023

The third public meeting of the Committee on Research at Multiple Scales: A Vision for Continental Scale Biology was held virtually. This was the second of three information-gathering sessions held via a webinar series.

Paving the Way for Continental Scale Biology: Tools and Approaches for Connecting Research Across Scales Agenda
June 15, 2023
12:30 p.m.–5:00 p.m.

12:30   **Welcome and Opening Remarks**
Jianguo Liu, *Committee Chair, National Academies of Sciences, Engineering, and Medicine*

12:35   **Panel 1—Applications of Observatory and Data Gathering Networks**
Lisette de Senerpont Domi, *Global Lake Ecological Observatory Network (GLEON), Netherlands Institute of Ecology*
Paula Mabee, *National Ecological Observatory Network (NEON)*
Osvaldo Sala, *Long Term Ecological Research Network (LTER), Arizona State University*
Margaret Torn, *Ameriflux, Lawrence Berkeley National Laboratory*

1:45   **Panel 2—Applications of Analytical and Sampling Tools**
Matthew Barnes, *Texas Tech University*
Jessica Ernakovich, *University of New Hampshire*
Susan Trumbore, *Max Planck Institute, University of California, Irvine*

2:55   **Panel 3—Applications of Data Integration and Artificial Intelligence**
Jennifer Balch, *Environmental Data Science Innovation & Inclusion Lab (ESIIL), University of Colorado Boulder*
Tanya Berger-Wolf, *Ohio State University*
Matthew Jones, *National Center for Ecological Analysis and Synthesis (NCEAS)*
Yaxing Wei, *Oak Ridge National Laboratory*

4:00 **Panel 4—The Role of Biological and Ecological Theory**
Elena Litchman, *Carnegie Institution for Science*
Alan Hastings, *University of California, Davis*
James Heffernan, *Duke University*
Christopher Kempes, *Santa Fe Institute*

5:00 **Adjourn Open Session**

AUGUST 21, 2023

The fourth public meeting of the Committee on Research at Multiple Scales: A Vision for Continental Scale Biology was held in-person. This was the third of three information-gathering sessions held via a webinar series.

The Keck Center, 500 Fifth Street, NW
Washington, DC 20001

**Paving the Way for Continental Scale Biology:
Technology, Techniques, and Teamwork for Connecting
Research Across Scales Agenda**
August 21, 2023
9:00 a.m.–4:30 p.m.

9:00 **Welcome and Opening Remarks**
Jianguo Liu, *Committee Chair, National Academies of Sciences, Engineering, and Medicine*

9:05 **Panel 1—Coordinated Data Collection & Theories**
Gillian Bowser, *Colorado State University*
Theresa Crimmins, *National Phenology Network*
Christopher Lepczyk, *Auburn University*
Daniel Park, *Purdue University*

10:05 **Panel 2—Indigenous Perspectives: Collaborative Approaches to Continental Scale Biology**
Stephanie Russo Carroll, *University of Arizona*
Cristina Eisenberg, *Oregon State University*
Danielle Ignace, *The University of British Columbia*

11:15 **Panel 3—Inclusive Training and Workforce Development: Promoting Diversity, Equity, and Inclusion (DEI)**
John Matsui, *University of California, Berkeley*
Bonnie McGill, *American Farmland Trust*
Milton Newberry III, *Bucknell University*
Sara Bombaci, *Colorado State University*

1:15   **Panel 4—Tools, Technology, and Research Techniques: Top-down Approaches**
Charuleka Varadharajan, *Lawrence Berkeley National Laboratory*
John Bargar, *Pacific Northwest National Laboratory*
Rachel Buxton, *Carleton University*
Sarah Huebner, *University of Minnesota*

2:15   **Panel 5—Tools, Technology, and Research Techniques: Bottom-up Approaches**
Nico Franz, *Arizona State University*
Jesús Pinto-Ledezma, *University of Minnesota*
Christine Wilkinson, *University of California, Berkeley*
Elise Zipkin, *Michigan State University*

3:25   **Panel 6—Across Sectors: Interdisciplinary Tools and Theory**
Brook Nunn, *University of Washington*
Andrew Farnsworth, *Cornell University*
Patrick Meyfroidt, *Université catholique de Louvain*

4:25   **Closing Remarks**
Jianguo Liu, *Committee Chair, National Academies of Sciences, Engineering, and Medicine*

4:30   **Adjourn Open Session**

NOVEMBER 13, 2023

The fifth public meeting of the Committee on Research at Multiple Scales: A Vision for Continental Scale Biology was held virtually.

Open Session Agenda
November 13, 2023
11:00 a.m.–11:30 a.m.

11:00   **Presentation on National Science Foundation's Technology, Innovation, and Partnerships (TIP) Directorate**
Erwin Gianchandani, *National Science Foundation*

11:30   **Adjourn Open Session**